The Dynamics of Spiral
PLANETARY MOTION

----

*A Revolutionary View of the
Spiral Governing Planetary
Motion in the Solar System*

By A.E. Dover

REVISED EDITION MARCH 2020

1ˢᵗ Edition kindle 2014

2ⁿᵈ Edition – print 2016

3ʳᵈ Revised edition 2020

Copyright © 2016 A.E. Dover.

All rights reserved.

ISBN:

ISBN-13:

The 2020 revision of this work seemed necessary to correct a number of errors and in order to improve and simplify the general outline of the subject, directed as it is, to astronomers professional and amateur. Also, to present in simple terms to the layman, the vital general principle of spiral motion evolving this wonderful Solar System.

DEDICATION

For a New Astronomy

----

*As Moths about a flame are doomed,
to be by that flame consumed.
See Planets about the Sun, gracefully gliding,
drawing ever closer,
with nature's spiral guiding.*

A. Dover.

HOW HIGH THE MOON?

Does our bright companion in space
move in fixed orbit about Earth in perpetual place?
Or, is either approaching or receding in headlong race?
Nature shows, in examples by the million, her law,
Nothing in the universe can remain, for evermore.
So we immediately toss out that fixed notion,
As science has done, for it's really 'perpetual motion'
Two possibilities remain for us to explore,
The theory that she is receding is the one to deplore,
Why we should do so, please let me explain.
Compare love between man and woman, the pull is quite plain,
As with all things in nature their behaviour is the same.
When two things are attracted they'll not grow farther apart,
But move ever closer, right from the start.
See, mighty oceans, rising to greet,
Old faithful Moon, clearly striving to meet.
Terra-firma moves likewise beneath our very feet
And surely you'll grant, t'is Mother Earth's very heart beat.
Isn't every tide, sweeping over rock and sand
Like the gentle caress of a true lovers hand?
Could they but speak, they'd tell helplessly, their destiny is planned.
Each companions nearness excites tremors to run deep,
As the nearness of lovers makes hearts 'miss a beat'
And nervously falter, hardly able to speak.

A. Dover. 1967.

PREFACE

Dialectical and Historical Materialist Concepts of Planetary Motion

Dialectical and historical materialism is the modern philosophy that recognises change in all things, transition of things into their opposite as positive does to negative.

My own curiosity initiated an interest in nature when, as a child, I was fascinated by water spiralling down the kitchen sink and how it stubbornly refused to go down in any other way. The next simple plaything was to rotate half a bottle of sour milk and observed that it generated a spiral motion and if I rotated it long enough globules of fatty substance began to form. Although it did not mean much to me at the time, my curiosity into matter and motion was aroused, a curiosity into spiral motion that has lasted all my life. In my teens I continually became aware of the spiral factor in nature and reflected in the sciences.

It was in the early 1960s that my interest in the subject of astronomy and in particular planetary motion, grew from my doubts regarding the claims by astronomy in general, that the Moon was rotating on its axis and that the period of rotation was described as synchronous with its orbital period. Also it was claimed that the Moon was moving away from the Earth. Knowing little of astronomy at the outset, yet instinctively I could not accept these claims, considering them as the result of mechanical thinking contradicting the general movement of our planets and satellites moving directly and indirectly toward the Sun. Two scientific principles had driven me in my approach to

Preface

planetary motion from my earliest perception. They were, first that there is no such thing in nature as a straight line, any line is part of a curve, any curve can be part of a circle as in mechanics or part of a spiral cycle in nature; secondly, we recognise that there is no such thing in nature, as a circle. That would be perpetual motion, straight lines and circular motion only exists in mechanics and mechanical thinking.

I was intrigued also by the general question of matter and motion and the fundamental question "basic forms of matter" raised by Frederick Engels in his work 'Dialectics of Nature'. However, my studies were intermittent and unorganised over many years, although I gradually progressed researching in subjects, astronomy and matter and motion. I was collecting a small mass of data and felt that I would have to continue research and publish my views eventually. By the 1980s firm ideas had matured in my mind concerning the sphere which was basic to both fields of my endeavour. It was then that my path of research in astronomy, narrowed down to spherical motion. My research notes I later collected and produced the work 'On Matter and Motion' (2005 unpublished), this was the precursor to 'On Planetary Motion' (2011 unpublished) and now the publication of this current concluding work. is the result.

My studies of planetary motion crossed paths with A.W.Drayson's studies into the 'Second Rotation of the Earth' (precession). As a consequence I spent several years closely studying his work which though based on the mechanical concept of the second rotation of earth being a circle, he has cleverly exposed in geometric detail, the misconceptions in astronomy concerning 'The Supposed Acceleration of the Moon's Mean Motion' and similarly the misconceptions concerning 'The Supposed Proper Motion of the fixed Stars'.

It was through Drayson that I focussed on, and realised the real significance of precession of the equinoxes as one of the two essential manifestations of planetary motion, the particular motion of the sphere itself, its Spin. I owe much to Drayson's questioning the cause of the obliquity of the ecliptic and his quite revolutionary endeavours and explanations with regard to the stars and the moon, all of which had been rejected by the authorities of his day. [1]

Having completed a work on the simplest form of 'matter and motion' I proceeded to examine the motion of the sphere. In doing so I naturally become immediately aware of two inseparable manifestations of the motion of the 'sphere', for embodied in the sphere is a dual motion, vis. its General motion and its Particular motion, more commonly identified as linear motion and spin. Although having identified matter in its simplest form, reducible no further than to the distinct and singular sphere, with no dual attributes to its material form, notwithstanding its transition from a mass of particles into a relatively solid spherical body; in the case of its motion, we find two distinct functional movements, the general and the particular, combined.

Perhaps we could describe the two aspects of the motion of the sphere combined, as being both an individual motion of its own and a general movement with and for others.

---

[1] The Cause of the Supposed Proper Motion of the Fixed Stars'. Lieut.—Col. Drayson, R.A. F.R.A.S. – 1874. (Sourced, British Library. 1n 1982.)

Preface

Accepting the dual nature of this simplest form of motion as a fact, we can identify that general and particular motion are also the inherent attributes of all higher forms of matter in motion. Here we progress to an understanding of the dialectical nature of the sphere and its motion, breaking right away from any fixed mechanical notions of the sphere and its motion as often applied to such as planetary motion; It became obvious to me that the Spiral was the essence of planetary motion, the mode of existence of the sphere.

I committed myself to an investigation of the two aspects, of spherical motion still bearing in mind the plan to follow the motion of the simplest form of matter in nature, the sphere; and what better than to let a general knowledge of the Solar system and the motion of the planets be my laboratory and guide, to assist me in my observations, as certainly no finer examples exist for the purpose of the study of the spiral and spherical motion.

Due to the nature of the subject my intention was to present a paper of my observations to academic organisations popular astronomy, and astronomical societies around the world to generate interest but after many attempts finding no interest in the subject over the years I have reluctantly adopted the publication process.

Alan Dover.

CONTENTS

Dedication iii

How High the Moon? iv

Preface v

Chapter I – Introduction 1

Chapter II – The Wonder of the Spiral in Planetary Motion 8
    The Two Fundamental Forms of Motion of a Sphere 9
    Circle or Spiral – the Mode of Earth's Orbit? 10

Chapter III – Earth's Spiral Orbit – Its Main Form of Motion 12
    Progression of the Pole and Plane of the Ecliptic 13
    The Rotation of the Line of Apsides 16
    The Cause of the Elliptical Orbits of the Spheres 18
    Kepler's Laws 19

Chapter IV – Earth's Spiral Spin 24
    Isaac Newton - Precession of the Equinoxes 24
    Newton's Serious Omission 27
    The Mode of Motion of all Spheres is the Spiral 29

Chapter V – The Solar System 32
    The Life Cycle of a Planet is Governed by Spiral Motion 32
    The Planets 35
    The Satellites 39
    The Accretion Process 42

Contents

Chapter VI – A General View of Earth Orbiting the Sun 47
Transition of Planets From Retrograde to Prograde Orbit 47
The Life Cycle of a Planet is Governed by Spiral Motion 58

Chapter VII – The Planetary Preconditions for Life on Earth 61
The Unique Conditions – The Earth Moon Union. 63
Geology –Stratification of Earth.. 68

Chapter VIII – A Web of Deceit — The 'Invariable' Solar System.. 72
The Newton — Hooke Correspondence 72
The Distortions of Newton, Laplace and Newcomb. 79
Isaac Newton on Orbital Motion 80
Newton on Diurnal Motion 83
P. Laplace 83
Simon Newcomb 89
Dr. R Hooke's Legacy 91

Chapter IX – The IAU – Perpetuation of The Deceit 93

References 96

CHAPTER 1
INTRODUCTION

    This approach to the investigation of planetary motion does not rest merely on observation and pure logic, it is firmly founded on scientific principles and the scientific data acquired over two thousand years concerning the dual motion of our planet, Earth. Today, This investigation lends itself well to the layman interested in astronomy since it requires no special mathematical talents to understand this natural phenomena. A knowledge of Plane geometry is all that is required for the layman to understand the basics of planetary motion.

    All one needs to understand the motion of the planets of the solar system are really rudimentary, plane geometry will suffice, based on the 360° points of the compass. and the Degrees Minutes and Seconds of arc (0° 0'0") of a circle. The whole research is to determine the measurement of planes and angles of the moving spheres. Each sphere in motion has two features, it has an axis of its spin and therefore possesses an equatorial plane, 90° to the axis. Each sphere also has an orbital plane about the Sun, Earth's is called the Ecliptic, it is customary to project a perpendicular line 90° upon the Earth's ecliptic and use it as a pole or axis of the ecliptic for convenience but of course, as an orbit it possesses no real axis like the spinning sphere produces. The Sun is the dominant sphere in the system and the movement of all the planets is governed by the Sun's axis, equatorial plane and even by the direction of the Sun's rotation which has an effect on the direction of rotation of the subject planets.

It is customary, for convenience in calculations for astronomers to relate angles of planes and axes and motion of the planets to the Earth's plane of the ecliptic. However, for our convenience we use the Sun's equatorial plane and axis, to examine the planets in their correct relation to each other and the Sun. The annual change in the angles of axes and planes of any planet in relation to the Sun, is the true measure of planetary motion.

The calculations requiring higher maths and spherical trigonometry have all been done over two thousand years by our masters who left us with annual record of the two motions of our sphere, Earth. The planet Earth is our focus and the issue since Isaac Newton's time, has been one of interpretation of those two annually recorded motions, the rotation the of Earth on its axis and its orbit The basis of the study of planetary motion therefore rests on those annual records. These records have been traditionally issued annually in the Astronomical Almanacs under the description of 'precession' constants'.

Each of these two annual movements, display the dual spiral, transitional, nature of planetary motion in all its simplicity. It follows that for the purpose of the study, one has first to become familiar with the annual movement of each of these two; particular attention must be made to identify the motion of each as affected by the gravitational attraction of the Sun . In doing so you will become aware of gross misinterpretations in modern (since Newton's time) astronomy,.

For any reader who might find it helpful as an introduction to the subject, I include this brief history of mans' progress in the search for an understanding of planetary motion. The simplest course of advance is through the works of ancient philosophers and past masters of natural science and mathematics. The orbits of the planets from the outset were considered to be circles. The issue of contention was whether the planets and Sun was circling the Earth (geocentric) or the Earth and planets were circling the Sun (heliocentric). We follow the general features of advance of Western or European astronomy beginning with Pythagoras through to Kepler with notes on contentions in Newton's time.

Pythagoras 580 – 500 BC

Pythagoras founded a school of natural philosophy following his tutor Anaximander. Here some of the ideas of Anaximander were taught, among which, was the notion of perfect circular motion. The belief was that the planets were spheres moving in circles about Earth as their centre, with Earth also in motion.

Simply stated, the Pythagorean perception of planetary motion was of spheres moving about a common centre (Earth) in perfect circles, Geocentric.

Aristotle. 384 – 322 BC

Aristotle, a student of Plato, also founded his own school of natural philosophy in Athens. He studied natural phenomena without the use of mathematics. For Aristotle the Earth was the centre of the universe but contrary to Pythagoras only in that, he considered that the Earth was not in motion. There was little change in the actual perception of planetary motion as being of the spheres in circular orbits about a centre. No real advance on his predecessors on this subject.

Aristarchus 310 – 230 BC

We do see a big advance with Aristarchus. It was Aristarchus who first postulated that the sun was the centre of the Solar System. Based on his geometrical observations and mathematical calculation of the Sun Earth and Moon, his findings were that the solar System was heliocentric.

This was a gigantic step forward. It laid the basis for the use of the more advanced mathematics and instrumentation by medieval astronomers to follow (Copernicus), to study the true motion of the planets of the Solar system. Meanwhile, there was to arise a setback to this advance, a revival of the geocentric model of the Solar system by Ptolemy:

Ptolemy 85 – 65AD

Ptolemy of Alexandria revived the geocentric model with a complex mathematical system to describe the motion of the planets about Earth as a centre, all in off-centred circles. His notorious work is known as the Almagest.

The geocentric conception of planetary motion, reduced again to circles about Earth, held sway until the mid 15th century when the heliocentric model was eventually restored by Copernicus.

Chapter 1 – Introduction

Copernicus. 1473 – 1543

Mathematics was being firmly established in astronomy through such as Copernicus who was both mathematician and astronomer. Mathematics had revealed the flaws in the geocentric Ptolemaic model. The Copernican model - De Revolutionibus Orbium Celestium was published in 1543. The heliocentric model had become the basis of the science of astronomy.

Though the heliocentric motion of the planets was restored the conception was still of circular orbits. We find also, that geocentric notions were not completely dead, revived again by Tycho Brae,

Tycho Brahe, 1546 – 1601

This Danish astronomer conjured up a geocentric model in which the Sun orbits the Earth and the planets orbit the Sun. This was met with little acceptance. But Brahe's contribution on the trail of discovery of planetary motion comes about through his skill in observations and instrument developments that he had passed on to his student Johannes Kepler who was engaged in studying the motion of the planet Mars. This was to lead to the second great advance in man's perception of planetary motion. Mention is first made of Galieo's role in confirming the heliocentric model.

Galileo Galilei, 1564 – 1642

Due credit is given for the great discoveries and contributions to the science of astronomy in general by Galileo Galilei , to his development of the telescope in particular, and his application of mathematics to the science. Galileo's observations led him to conclude that the Copernican heliocentric model of the Solar System was superior to the Ptolemaic model. Mathematics, he considered, is the true language of science, his 'Dialogues Concerning the Two Chief World Systems' were published in 1632.

The heliocentric model of the Solar system now held fast but planetary motion was still viewed as consisting of circles, as a result, Kepler was to be in good stead when he emerged from the shadow of such predecessors to advance further the understanding of planetary motion.

Johannes Kepler. 1571 – 1630

Kepler, a mathematician, and a student of Tycho Brahe became skilled in observing the orbits of the planets. As a result he was to eventually propound the 3 laws of planetary motion. He derived his three Laws from observations, in particular of Mars; they can be re-stated as :

The orbits are ellipses with the centre of mass at one focus if the two bodies revolve about each other under the influence of a central force, the line joining them, sweeps out equal areas in equal times.

The square of the period of the orbit is proportional to the cube of its semi- major axis

Calculating the exact position of a planet in the solar system at any given time was now possible. The third law is the basis of the law of gravitation, the mathematical formula to be further developed by Isaac Newton as the law of gravitation. The theoretical basis for the advance into the much later space age was laid. Since the time of Kepler, the science of Astronomy has no longer been concerned with planetary motion as such, what more was needed to know planetary motion was now seen as being elliptical orbits? The orbital position of the planets was calculable, now the science could move forward on that knowledge and was content. Astronomy turned to focus attention on the physical properties of the planets rather than a closer study of their motion and looked outward to the stars and deep space. Today, the knowledge gained i.e. the ability to calculate the exact position of a planet in the solar system at any given time, now serves the political and economic powers of the day. That is enough it seems.

This is how the situation stands today in the science of astronomy, with regard to the lack of understanding of planetary motion, remaining as an impediment to the understanding of the unique temporary planetary conditions that have made life on Earth possible. My work is an attempt to advance and complete our perception of this wondrous natural phenomenon, planetary motion that creates the conditions for life, continually influences life and ultimately destroys life.

From Kepler's closed elliptical orbit we begin our quest with the concept that a planetary orbit is not merely an elliptic circle but is an elliptical spiral orbit, being of infinite difference, with consequences of great interest to the future of astronomy and geology. So let us take the spiral by the tail and let it lead us through the life cycle of all planets and satellites.

CHAPTER II

THE WONDER OF THE SPIRAL IN PLANETARY MOTION

> The Mode of Existence of the Sphere is its Particular Motion, its Axial Rotation. Its Destiny, is Governed by its General motion, its Orbit about its primary.

There are two movements of our planet that are recorded annually. The first is its spin as affected by Sun's gravitation producing (precession) together with the Moon's gravitation (nutation) it is known as Luni - solar precession, they produce the annual shift of the equinoxes as explained in the Almanacs measured annually as the Earth's equatorial plane shifts in a retrograde direction along the ecliptic at 50.2", quite a correct explanation as advanced by Newton.

The second movement of Earth records the movement of our Earth's annual orbit A prograde movement, that is similarly recorded as the ecliptic crosses the equatorial plane on its annual orbit, revealing a shift of 0.1247" a much smaller amount by comparison with Spin motion.

These are the two vital measurements of annual motions recording Earth's orbit and Spin, as can be seen in the statements in the Almanacs. However, this prograde, annual shift of the Earth's orbit of 0.1247" along the

Chapter II – The Wonder of the Spiral in Planetary Motion

equator is officially credited to planetary perturbation causes. (This is an incorrect observation negating the Sun's as the true cause of the 0".1247 shift of the ecliptic)

The Two Fundamental Forms of Motion of a Sphere

We need to understand that these are the two fundamental forms of motion inherent in all spheres, spin and orbit. We also need to understand how the spiral dynamic effects each form differently. It is only when we have done this is it possible to understand the true motion of our planet and indeed the whole solar system.

Also, we can then reveal the history of how the existing misconceptions and mystery that arose and permeated the Science of astronomy for the last 360 years. The whole subject rests on the influence of the Sun's gravitational attraction, the two fundamental motions of our planet, its spiral spin and orbit

It is strange in this day and age that the motion of the planets in the Solar system is said to be based on them orbiting perpetually in elliptical circles. Any suggestion that this is not so, is strongly rejected by astronomy in general.

350 years have passed and still our science is locked into this insidious medieval notion. This then is the story of how and why the purveyors of the corruption of a science maintain the farce to this day.

The case for the notion of the Solar system being constantly changing, due to being systematically governed by the dynamism of the spiral orbit and spin, emerged in 1679 when Dr, Robert Hooke approached Sir Isaac Newton for his opinion of his recently published observations on planetary motion. A short series of correspondence and discussion ensued culminating with Dr. Hooke explaining the concept of the elliptical spiral as the form of planetary motion, clearly defined in a sketch.

This was the point at which the liberation of modern astronomy from the medieval stagnation of the science was commencing, though to be stalled it had arrived in man's conceptions.

Circle or Spiral – the Mode of Earth's Orbit?

Since Kepler's fine contribution defining the elliptical nature of planetary orbits to science, no additional advance has since been made concerning planetary motion. The reason for this is the stubborn refusal of astronomy to recognise the spiral nature of all planetary motion. Thus, any advance in the understanding of orbital or rotational planetary motion has not advanced, holding the science to medieval notions concerning the Solar System.

In considering the Solar system, a system of spheres, from the dialectical-materialist point of view, our first observation is that nature produces no perfect circle in planetary motion or straight line – such would be perpetual motion. The law of gravity operating in the Solar system is absolute, drawing all satellites by means of spiral orbital paths to their primaries and all planets to the Sun. From the above arises the materialist theory of planetary motion in contrast to the current mechanical-materialist viewpoint that ignores, or pays little heed to the above criteria when it suits their metaphysical arguments.

References to the mode of Earth's orbit about the Sun is found throughout all scientific astronomical media, including the IAU. dictates that the orbit is a circle; this forms the official international physical definition. An 'elliptic circle', a closed circle no less. A random sample follows:

> "The Earth circles the Sun in a flat plane. It is as if the spinning Earth is also rolling around the edge of a giant, flat plate, with the Sun in the center. The shape of the Earth's orbit, the plate changes from a nearly perfect circle to an oval shape on a 100,000-year cycle (eccentricity)."

And how is this Wikipedia definition for an amusing piece of sophistry, currently circulating, "a non repeating trajectory" and a "regularly repeating trajectory" each are pseudonyms for a circle and a spiral orbit.. You can choose which you prefer it seems: [2]

> "Normally, orbit refers to a regularly repeating trajectory, although it may also refer to a non-repeating trajectory. To a close approximation, planets and satellites follow elliptic orbits, with the central mass being orbited at a focal point of the ellipse, as described by Kepler's laws of planetary motion.

Yet, also in the same source, it speaks of a planetary body: that due to drag *'will rapidly spiral down and intersect the central body.'*

There is also to be found common variations on the theme, such as those that unwittingly contradict the circular orbit dogma by claiming that a planet like Earth or a satellite like our Moon, is orbiting away from, or moving closer to its primary. They seem unaware that either of such orbits they describe can only move in a spiral fashion. It is almost recognition of the spiral in motion.

---

[2] https://en.wikipedia.org/wiki/Orbit

CHAPTER III

EARTH'S SPIRAL ORBIT – ITS MAIN FORM OF MOTION

The effect of our Sun's gravitation on Earth's orbit is simpler than its effect on its rotation, so we will examine our spiral orbit first

We accept that the underlying principle in the relationship between the plane and axis of a rotating body is that the two are part of the same movement. The axis arises with the rotation and always stands perpendicular (90 deg.) to the plane. If the equatorial plane moves one degree so does the axis move one degree in unison.

Fig. 1

The principle underlying the relationship between the axis and plane of a rotating body is, that the angle of the axis, always remains perpendicular to the plane, 90°.

Completely

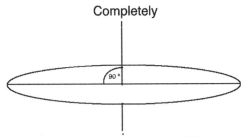

In the case of astronomy's evaluation of the precession of the equinoxes, initially by Sir Isaac Newton; the above principal was completely disregarded.

Chapter III – Earth's Spiral Orbit – Its Main Form of Motion

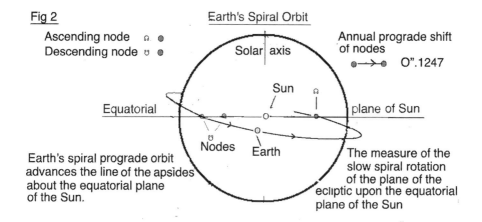

Fig 2 Earth's Spiral Orbit

Ascending node ☊ ●
Descending node ☋ ●

Solar axis

Annual prograde shift of nodes
●→● 0".1247

Sun

Equatorial   plane of Sun

Nodes Earth

Earth's spiral prograde orbit advances the line of the apsides about the equatorial plane of the Sun.

The measure of the slow spiral rotation of the plane of the ecliptic upon the equatorial plane of the Sun

Above we view the Earth in its prograde spiral orbit. the Ecliptic and we can see how the Earth crosses the equatorial plane of the Sun at the nodes, so annually advancing the ecliptic along the circumference of the equatorial plane of the Sun. This is a very slow cycling, spiral movement of the ecliptic about the Sun, annually at 0"1247.

This is consistent with the retrograde movement of the equinoxes, both spin and orbit influenced by the same impetus from the Sun.

Progression of the Pole and Plane of the Ecliptic

Fig. 2 Below, again shows the dual dynamics of the spiral orbit, the path of the Earth's orbit, measured annually as it moves around the equatorial plane. This acts together with the second aspect, its inherent movement to the centre; Earth's path is changing the angle of obliquity, by moving closer to the Sun in one smooth function. Each complete revolution of the ecliptic about the Sun, due to the spiral, will minutely decrease the angle with the Sun, drawing closer to the Sun.

The Dynamics of Spiral Planetary Motion

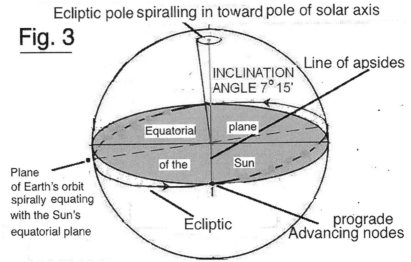

Ascending node of Earth's path has annual prograde shift of 0".1247 due to spiral progression.

Progression of the Pole and Plane of the Ecliptic

Accepting that the spiral law governs all planetary motion, as it does, then the picture Fig. 2 comes to life. To explain, the angle shown is the present angle between the ecliptic and the equatorial plane of the Sun which is 7' 15". The prograde spiral motion of the nodes, at the junction of the ecliptic with equatorial plane of the Sun moving along at 12".47 a century will eventually execute a complete circuit. Because this is a spiral movement also moving towards a centre, it will annually reduce the angle between the ecliptic and the equatorial plane of the Sun.

What you see here is part of the life cycle of not only our planet but all planets in the Solar system. The Spiral law is the key to planetary motion.

As current astronomy would have us believe, the Earth in its passage about the Sun is performing closed elliptical circuits on a flat plane, being merely subject to planetary

Chapter III – Earth's Spiral Orbit – Its Main Form of Motion

perturbations. This is a deliberate fallacy that contradicts the long - standing scientific criteria concerning nature and motion that acknowledges that there is no such thing as a closed circle and no such thing as a straight line.

The true definition that has not been acknowledged, is that in nature the spiral is the mode of existence of all spherical matter in motion, only in man's science of mechanics does the straight line and circle exist. We understand that the Earth's orbit is following the course of an elliptic spiral about the Sun

The mean plane of the ecliptic is actually contributing to bring its pole and plane to equate with the equatorial plane and pole of the Sun.

Astronomers have not recognised this *spiral* motion of the Earth's orbit for what it really is, the real cause of the annual 0".1247 shift; just as it operates for Earth's spin.

It is, the most obvious cause of the rotation of the mean plane of the Ecliptic and the Line of nodes; being the measurement of the motion of Earth's second form of motion, its orbit.

The science has chosen to attribute the cause of this most natural motion of the plane of the ecliptic, to the gravitational influence of our surrounding sister planets and, consequently, have named the 0".1247 annual movement "planetary precession".

In **Fig. 2** we can see how the annual spiral of the ecliptic changes the inclination of the ecliptic to the Sun's equatorial plane and the ascending node of Earth's path records an annual prograde shift of 0".1247 deg. the component affecting general precession. There is no such thing as a closed ellipse in planetary motion. Like all other aspects of planetary motion, the rotation of the Line of the Nodes will not be performing a closed ellipse. It takes a great number of orbits of Earth about the Sun to bring about one single complete rotation of Earth's Line of Apsides about the Sun as depicted in Fig. 4.2. (greatly exaggerated for clarity of the slow rotation of the ecliptic about the Sun).

The Rotation of the Line of Apsides

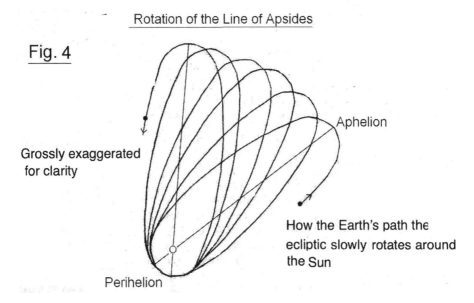

Currently the detected movement of the plane and pole of the Ecliptic, is viewed in a mechanical way, as being a disturbed path of a closed elliptical revolution, not as part of a spiral resulting from the progressive movement of the orbital path of Earth.

It is not at all surprising that other planets are found to exhibit 'precession' movements, recently, Mars for instance. The fact can no longer be obscured that precession and progression are essential components of planetary motion, vital evidence revealing that the spiral is its living form.

We bear in mind also, that the 'precession' of the ecliptic is not a true 'precession' movement; by reason that the pole of the ecliptic is not the pole of an axis of a spinning sphere but merely a line drawn perpendicular to the ecliptic. Only a spinning sphere can produce an axis performing a precession movement. In addition, it has to be in a retrograde motion to its primary. The plane of the ecliptic and its pole are prograde.

As we note in **Fig. 2,** the pole of the ecliptic will be tracing a slow spiral on the celestial sphere, and its course is in the same direction as the orbital path. Not in the opposite direction like the precession axis of a spinning sphere, a true precession movement. Thus. there is a case for altogether redefining what is, or was known as "planetary precession".

On the proper understanding of the movement of our planet, that is, of the dual attributes of its motion, its particular (spin) and its general (orbital) motion related to the historical annual records of those annual amounts. The way is open to calculate with some degree of accuracy: To calculate and approximate how many years until that point when the Earth's axis and the pole of the ecliptic will stand perpendicular to the equatorial plane of the Sun. When the Sun will daily rise in the East above the equator then passing along and above our equator to set in the West every year. Our annual seasonal changes will have gradually diminished, to disappear altogether at this point and the Earth continually slowing may well no longer be spinning or possess an axis; though it will still be orbiting closer, transfixed as is the Moon at present orbiting Earth.

The Cause of the Elliptical Orbits of the Spheres

The first aspect of spherical motion to consider, is the General or linear motion of the sphere a movement about a centre, a spiral in fact, a very simple but dual motion conforming to strict laws, it being understood that there is no such thing as a straight line or perfect circle to be observed in the motion of matter; they have no place in nature, as all is accomplished with the continuous curve, the spiral. So, firstly for the spheres of the Solar system all linear motion is orbital, about and toward a centre.

The spiral movement of a body can be observed in either one of two directions, relative to its primary, theoretically in a clockwise or an anti-clockwise i.e., prograde or retrograde direction. Though in the case of the Solar system all planets orbit in an anti-clockwise direction. The dynamics of general motion are best explained and advanced from Kepler's second law of planetary motion describing the radius vector sweeping equal areas in equal times. Fig.2.3 The orbital velocity varies by increasing and then decreasing in accord with the dynamics of the elliptical orbit from which the mean velocity is derived. The Earth is depicted travelling in anti-clockwise direction increasing in velocity from A to B and decreasing in velocity from B to A. In each orbit a satellite responds to the law of gravity between it and its primary that overcame its independent path to the Sun and ultimately leads to its complete loss of independent existence becoming interdependent on and to subsequently merge with its primary.

Chapter III — Earth's Spiral Orbit — Its Main Form of Motion

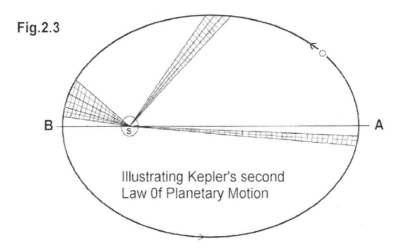

Fig.2.3

Illustrating Kepler's second Law Of Planetary Motion

Kepler's great contribution to the solution of the problems concerning the understanding of planetary motion was a big step forward for astronomy. His three laws of planetary motion paved the way for great advances by creating the ability to calculate and predict the movements and position of all the bodies in the solar system. Although Isaac Newton followed with further mathematical calculation in developing related aspects. Our starting point to investigate further the general motion of the sphere, continues from J. Kepler's laws of planetary motion which describe the universal elliptical shape of all orbits during the greater part of a satellite's existence :

Kepler's Laws

1.) His first law simply stated is that the planets orbits are elliptical with the centre of the sun at one focus.

2.) His second law concerns the radius vector sweeping equal areas of the elliptical orbit in equal times. Which demonstrates the varying angular velocity of an orbiting planet from aphelion to perihelion and then from perihelion to aphelion, accelerating to the one and decelerating to the other.

3.) The ratio of the squares of the periods of any two planets is equal to the ratio of the cubes of their average distances from the sun.

We recall as shown by Galileo and later explained by Newton in the first of his three laws of motion :

1). Every particle continues in a state of rest or motion with constant speed in a straight line unless compelled by a force to change that state.

This is indeed the case for all matter within our solar system the third law states

3). All forces arise from the mutual interaction of particles and in every such interaction the force exerted by the one particle on the second is equal and opposite to the force exerted by the second on the first, or as it is usually expressed: action and reaction are equal and opposite.

The general theory accepted by science concerning the evolution of planets is that a general motion of particles within a gaseous mass moving about and toward a centre (accretion) brings into being particular spheres. Such is the evolutionary theory of the Solar system.

We find no general motion of spheres moving or tending away from a centre in the Solar system, all movement is gravitating to a centre. That is not to say that matter in spherical motion, does not or could not move in orbital manner, away from a centre in another magnitude, such as stellar or Chemistry for example.

Chapter III – Earth's Spiral Orbit – Its Main Form of Motion

The main thrust of Kepler's contribution was to move man's notion of planetary orbits from circles to ellipses and this established the basis for calculations based on the dynamics of an elliptical orbit. Kepler's work was the basis for the next step to advance our understanding of planetary motion, that is, the recognition of the spiral component of planetary motion, the essence of the life cycle of the spheres. whether planet or satellite in both their orbit and spin, their general and particular motion in the Solar system of accretion.

I was fortunate to discover long ago in the 60s, through my own research and observations, that the sphere was the simplest form of matter; this was in order to derive what I had initially sought, the simplest form of motion; Since there is only matter in motion, motion can only be identified in forms of matter; the sphere it seemed to me, was the simplest form that matter can take, hence, it became my objective to study its motion which is of course the simplest form of motion. A study of spherical motion in the solar system led me to the obvious conclusion of which the spiral is the essence of both orbital motion and spin. I display the orbital motion of the sphere, its general, planetary motion in Fig.2.4, My investigation into the several other aspects of spiral planetary motion soon followed.

Fig.2.4

The path of a satellite 'T' entering into elliptical orbit about a primary 'S'

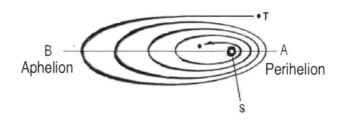

Here, I have depicted the general motion, of capture, submission and demise of a satellite 'T' to its primary 'S'. It explains simply why an orbit is an ellipse. In the very first instance the gravitational influence of the primary body 'S' will distort the path of a free, passing body without capture. But if the mass and proximity of the passing body be such that the distortion to its path becomes a curve, centred on the primary, decreasing in velocity in its progress up to a point 'B' (aphelion) distant from the attracting body 'S' Then, being unable to continue on its own independent path to escape from the influence of the attracting body, it now proceeds along a closing path with its primary from aphelion to perihelion 'A' increasing in velocity. Once captured in this initial orbit, then the path of the once independent passing body takes on the elliptical characteristic and properties of a satellite obeying the elliptical law of 'equal areas in equal times' producing, albeit in the first instance, the most pronounced elliptical path of a satellite about its .primary. In this condition the now orbital path least resembles a circular orbit. Yet we find, through acknowledging the spiral as a factor of celestial mechanics and planetary motion, together with the law of 'equal areas in equal times' in elliptical orbits; a satellite continuing in this life cycle slowly draws ever closer to its primary. Thus a satellite will gradually lose its elliptical orbit, becoming a more circular one as it closes to join with its primary eventually orbiting about the equatorial plane of the primary. This latter influence of the primary over the satellite is achieved by means of a process of spiral orbital 'progression', (I take the liberty of using this term progression to clearly distinguish the movement from the customary misuse of the term 'precession' for the shift of the ecliptic). Few satellites are likely to have been captured orbiting precisely on the equatorial plane of the primary, or at their present inclinations as is currently, generally considered. All orbital motion progresses; the course of all the satellites of the planets from outer to inner about the Sun, indicate this to be an essential factor in the process of planetary motion.

This spiralling orbital process of formation, capture and demise displays the life cycle of a satellite from beginning to the end where it joins with the primary. All our planets are the end result of minor spheres becoming subject to the greater spheres, their greater gravitational influence governs the process of 'accretion' a process of planetary evolution. After all, all spheres in the Solar system are but satellites to the Sun There are now, no free bodies orbiting the Sun that are potential planetary satellites. The planets have, through the process of accretion, consumed almost all of the lesser spheres, leaving the few remaining captured satellites to receive the same fate, as all proceed directly and indirectly to the Sun.

CHAPTER IV

EARTH'S SPIRAL SPIN

Isaac Newton - Precession of the Equinoxes

> **PRECESSION**
>
> The equator and the ecliptic, and hence the equinox, are continually in motion. The motion of the equator, or of the celestial pole, is due to the gravitational action of the Sun and Moon on the equatorial bulge of the Earth; it consists of two components, one luni-solar precession being the smooth long-period motion of the mean pole of the equator round the pole of the ecliptic in a period of about 26,000 years, and the other nutation being a relatively short-period motion that carries the actual (or true) pole round the mean pole in a somewhat irregular curve, of amplitude about 9" and main period 18.6 years. The motion of the ecliptic, that is of the mean plane of the Earth's orbit, is due to the gravitational action of the planets on the Earth as a whole and consists of a slow rotation of the ecliptic about a slowly-moving diameter, the ascending node of the instantaneous position of the ecliptic on the immediately preceding position being in longtitude about 174deg. this motion is known as planetary precession and gives a precession of the equinoxes of 12" a century and a decrease of the obliquity of the ecliptic of about 47" a century.
>
> In this sub-section the effects of the motion of only the mean poles of the equator and the ecliptic, known as general precession, are considered; the effect of nutation is dealt with in sub-section C. The treatment is restricted to the development of formulae for the practical application of corrections to coordin-ates and orbital elements.
>
> (Explanatory supplement to the Astronomical Almanac. Circa 1982)

Chapter IV – Earth's Spiral Spin

A crucial aspect of the spinning Earth is the precession dynamic of its reaction to the attracting forces of the Sun and moon, causing precession of the equinoxes (as explained by Isaac Newton in 1687 in the Principia) responsible for an annual precession of the equinox of 50".2. This we recognise as the measurement of Earth's second form of motion, its Spin, responsible as Newton suggests, for the tendency to pull the equatorial plane to equate with the plane of the ecliptic and of course this, will in accord, annually spirally move the north pole toward the centre, the pole of the ecliptic. Thus the movement of the pole together with the equatorial plane is due to the Sun's gravitation. It is the spiral inward movement that produces the annual decrease in the Angle of obliquity.

The spiral component has two functions, one advancing the equinox along the plane of the ecliptic and two, the Earth's orbit moving to the centre, annually decreasing the angle of the obliquity by 0. 47". By ignoring this dual function it is not possible to properly understand the true cause of this decrease.

So astronomy had left itself without an explanation of the cause of the annual decrease in the obliquity. By ignoring the spiral component No one has pointed out or questioned this omission of Newton's and it has ever since been discreetly ignored by astronomers in general and replaced by the ridiculous Laplace planetary theory which attempts to conceal the fact that the annual decrease in the obliquity of the ecliptic is the product of the spiral nature of the precession movement just as Newton illustrated.

This is the crux of the matter concerning the annual decrease in the obliquity whereby, the science of astronomy became seriously corrupted. It is logical to believe as Newton demonstrated, that it is the rotation factor that is affected by the Sun's gravity on the equatorial bulge that is responsible for precession of the equinoxes.

It is important to note in the official descriptions, that it is in no way suggested that, *'Luni-solar precession'* is responsible for the pole of the equator moving closer to the pole of the ecliptic by any increment at all. Quite the opposite, this factor is credited to 'planetary precession' as being wholly responsible for the decrease of the obliquity of the ecliptic of 47" a century.

To proceed to really understand the laws of planetary motion, the error of this basic official description of the dynamics of precession must be clearly understood, since therein lays the concealment of a basic law of planetary-motion. The task here is to disperse the obscurity and the key to doing so, is to determine, what really is the cause of the annual decrease affecting the obliquity of the ecliptic?

The upshot of this theory is, that the course of the pole of daily rotation is considered to be describing a *'circle'* on a flat plane

The effect of the Sun on the prograde spinning body of Earth is the precession cycle, a peculiarity to spherical motion but subject to the spiral law of motion none the less. The science of astronomy lost the plot in the 17th century by ignoring the spiral motion of the planets, which was the next scientific step forward from Kepler's gift to them, the elliptical orbits of planetary motion. What follows is the story of the fallacy of the 'invariable' Solar System this is still perpetuated today.

Newton's Serious Omission

Sir Isaac Newton first identified the effect of the Sun and the Moon on its orbit as Luni-Solar precession, with a drawing displaying the function as below. It can be seen that he took account of the fact that the Sun was tending to pull the equatorial plane down to meet with the plane of the ecliptic. This was seen to account for the Precession of the equinoxes all were satisfied with this advance. But, here we draw your attention to a couple of points of omission in this description of precession. Deliberate executed or not, they are omissions.

Most important we note that he indicates that the plane of the equator is being drawn down to meet the ecliptic, the consequence of this movement must also draw the pole of Earth's axis to meet with the pole of ecliptic. This is elementary geometry, but this fundamental fact is ignored.

This decrease in obliquity is a factor affecting Earth's climate on an annual basis. Why has it been consistently ignored by astronomers? This actual decrease is cumulative being 47" of arc seconds per century.

According to Newton's graphic there is no decrease in the angle of obliquity or a natural decrease in the angle indicated between the equatorial plane and the Ecliptic. Obviously, If he had, he would have had to recognise the spiral. Seemingly no one has ever questioned it because they understood and new better, the 'invariable' solar system notion had to be preserved.

Here we view Newton's description of the cause of the precession of the equinoxes. The bending moment

The Dynamics of Spiral Planetary Motion

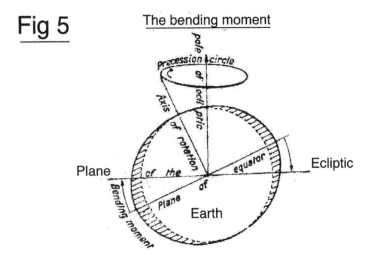

Fig 5

Precession. The exaggerated equatorial bulge of the Earth is shaded

The conclusion drawn from this formulae was the recognition of the annual precession movement of the equinoxes along the plane of the ecliptic. This produced the polar curve traced on the celestial sphere called the precession circle but there was no recognition of the 0"47 annual decrease of the angle of the equator to the ecliptic caused by the spiral

The result of ignoring the spiral mode of planetary motion in the precession cycle depicted the whole precession movement as a circle. Below, I present the Fig, of Newton's definition of the precession of the equinoxes. with the spiral interpretation added as it should have been. Under the influence of Sun's gravity upon a slowing rotating Earth, the inertial movement of the planet toward the centre can only take the form of a Spiral.

The movement toward the centre is not complete until Earth joins with the Sun. When the Earth is finally orbiting the Sun on its equatorial plane. The planet is then on the final trajectory for entry into the accretion process.

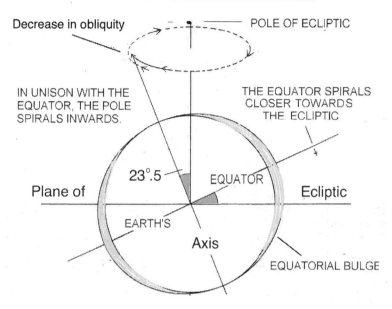

Fig 6 — Newton's Fig. updated to show the spiral with decrease in obliquity

The two outstanding facts we learn from our sciences is that in nature there is no such thing as a straight line and no such thing as a circle, such things belong to mechanics. Therefore, there is no such thing as perpetual motion all things are subject to change and the Spiral is the dynamism for change.

The Mode of Motion of all Spheres is the Spiral

In Fig.3, 3 is depicted the standard and popular but poor mechanical comparison of the dynamics of a spinning top used to explain the dynamics of precession but here too, we can show the spiral dynamic that exists affecting the axis of spin even in the mechanical model, but which is always overlooked. Principally, the motion of the top is not a 'wobble' as described but a continuous spiral movement in generate or degenerate motion. Its strongest use is the demonstration of the principle that the angle of the axis of the spinning body can only change through the spiral form of motion. A natural Law of all spherical motion.

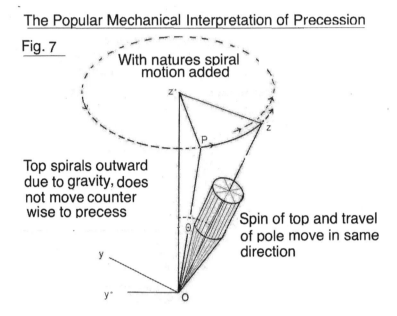

Fig. 7

The Popular Mechanical Interpretation of Precession

It will be realised that the dynamics of a spinning top, is under the influence of Earth's gravity and yet still obeys the spiral law of spherical motion. The pole P is not describing the curve of a circle along P – Z but the curving path of a spiral; a spiral path that is moving outward away from the O – Z° axis. The axis of any spinning body, such as the illustrated mechanical top cannot describe a circle, since that would be perpetual motion. A change in the angle θ indicates a spiral movement increasing or decreasing that angle. The idea of an axis performing a precession 'circle' is a mechanical materialist metaphysical concept that is counter-posed by the dialectical concept of a spiral that recognises change. Even in the case of minor increase or decrease of 1or 2 deg. in the angle, it could only be accomplished by a spiral movement of the plane and poles of the spinning body. Even in the mechanical Top.

The Earth's curved precession path is part of a spiral indicated by the annual decrease in the Obliquity of the Ecliptic, thus the polar movement of the Earth's axis is toward alignment with the pole of the ecliptic or better stated, the Earth's equatorial plane is becoming aligned with the ecliptic by means of a spiralling rotational motion in conformity with planetary motion. All changes in planetary motion are achieved by means of the spiral. If this be evidence of a law of planetary-motion which must be applicable to all spheres and more so to spheres that are not perfectly spherical, but oblate; then the same law applies to all such satellites as well as to primaries.

The conclusion in astronomy has been to see the precession circle as a "wobble" or oscillation with only 2-3 deg. effect on the decrease in the O/E. But there is no recognition of the spiral component that makes the axial motion possible. To describe, such a movement, due to any cause, without the spiral component, is impossible, a fiction.

CHAPTER V

THE SOLAR SYSTEM

The Life Cycle of a Planet is Governed by Spiral Motion

Astronomers have generally accepted that it is from a rotating disc of primordial matter that the Sun and our planets are formed. This view is evidenced by the discovery that there is a disc seen to be surrounding and presumably rotating about, 'all young Suns'. Such discs are thought to be the primordial mass from which the still forming Suns and their system of planets emerged.

The process has a beginning and an end, beginning with a primordial rotating mass of matter from which a centre is established, a Sun, to which all the matter within the mass is gravitating in a direct or indirect orbiting manner. The law of gravity in the realm of spheres can be seen to operate through a process of spiral motion, which is governed by the dynamics of the spheres two forms of motion, its rotation and its linear motion.

Commencing from this widely held and most likely theory of the spirally rotating disc of primordial matter, it seems that numerous smaller local congregations of matter, within this primordial disc, begin rotating with a spiral motion toward their own individual centres and subsequently each will form rotating accretions of matter

Chapter V — The Solar System

about their individual centres. It follows that, from the direction of the general spiral rotation of each of these local congregations, comes the establishment of the direction of rotation of the now relatively rapidly spinning spherical bodies forming at their centres. The planes and direction of rotation of these individual accretions would no doubt vary considerably, some may well be formed rotating on their axis in a retrograde direction with regard to the plane of orbit of the primordial mass. But all would be orbiting in their linear motion, in accord with the mass movement about the common centre the forming Sun.

The common factor in celestial-motion is the sphere and its motion, our solar system and its surrounds are a magnificent display of the formation, life, and demise of spheres, even from galaxies to Suns, to planets and their satellites; all obeying the same laws in the most beautiful demonstration of the simplest form of matter in motion, the sphere. Just as all the planets obey the laws of planetary motion, compelling them to ultimately orbit their primary, the Sun, through a process of accretion about its equatorial plane, so too are the satellites moving about the individual planets.

All planets in the solar system obey the fundamental law of gravitation which is the basis of the laws of planetary motion operating through planets performing a decreasingly elliptical and spiralling motion inwards, towards their primary the sun. At the same time, being induced to advance or progress their orbit to ultimately equate with the equatorial plane of their primary the Sun. Planet Earth is no exception to this law. All planets and satellites are induced to orbit in the same direction as their primary rotates about its axis. Previous chapters have dealt with this aspect, some attention will now be given to the spin of the spheres in the solar system. All the planets in a protracted process of accretion, are induced to rotate prograde and their equatorial planes are being induced through precession to equate with their orbital planes. Notable is the fact that to date, as in Kepler's laws, the perception of the laws of planetary motion are limited to the orbital (general) motion of the spheres and remains as a mechanical vision of closed ellipses. It is notable that there has been no proper investigation into planetary motion with regard to the rotation of spheres (particular motion). Particular-motion is an equal component of planetary-motion since it is part of the transitional sequence of all matter proceeding from a primordial mass through spheres to join with a star such as the Sun.

In fact, matter in general-motion is sequentially transformed into its opposite, particular-motion, as in the primordial mass of matter seen in general-motion within which begins numerous movements about a centre, each becoming an individual spherical body identified with its particular–motion, spin.

Eventually a satellite after losing its particular motion, its spin such as our moon and many other moons, begins its entry or absorption into its primary, by a process of breaking down. This will begin first on entering Roches limit, where the satellite begins breaking up into large and then smaller pieces and then eventually over a long period into particles and eventually (subject to atmospheric conditions about the primary) even heated into a gaseous state in a general-motion about its primary. This is a natural process prior to merging and becoming part of the particular-motion of the primary body. Thus, completing the process of transformation of a form of matter and motion into a higher form; of satellites combining to form a greater sphere, every stage of this process can be viewed in the Solar system laboratory. The transition of matter in the Solar-system, from beginning to end is continuous process, of moving from one spherical form to a higher spherical form, the mode of transition of this form of matter is, throughout, undoubtedly the spiral with its dual capabilities. Presumably a heated burn up entry for a satellite is always dependent on the primary having a substantial atmosphere of some composition.

As to the spin of the planets, they are subject to precession due to the gravitational attraction of the Sun on a spinning sphere, to equate its equatorial plane to first to the plane of the ecliptic and subsequently to the equatorial plane of the Sun

The Planets

Considering first the inner planets, Venus has a retrograde particular motion, slowly spinning at an obliquity to its orbit of 177 deg. which being in such proximity to the Sun and close to the end of its independent existence; does, with regard to its particular motion stand in contradiction to all of the other planets and satellites; Venus is a clear exception to the rule that requires an explanation. I consider that Venus is in the last stage of her polar rotation, 3 deg. after which she will be prograde. We know that the Earth's decrease in obliquity is influenced by its oblate circumference about its equator, causing Earth to progress its equatorial plane toward that of its orbital plane. Without this oblate feature, perhaps Earth too could not resolve any difference between its equatorial plane and its orbital plane. Perhaps the axial precession in particular motion is entirely dependent on the degree of oblate ness of the sphere although I would venture say that it is not totally dependent on the oblate ness.

Both of the inner planets Venus and Mercury, have presumably, long since absorbed any moons and who knows what original angle the orbital paths of Mercury and Venus formed, or how many moons they each held and consumed on their determined paths to the Sun. Mercury is all but (0° 15' 0".) fully aligned for her final descent into the Sun at her equatorial plane, likewise Venus with her orbital plane at approx. 3° is close behind.

With Earth's orbital plane at a mere 7° 15' angle to the Sun's equatorial plane, it is most likely that Earth has a history of accretion, of consuming several or a number of moons of varied size. Geology when properly oriented on the spiral process of planetary-motion may discover evidence of this. Fortunately we, in third position from the Sun, will have the opportunity to deduce the whole story once freed from mechanical materialist concepts of planetary-motion, for there, stretched before and behind us, is the complete story of planetary motion; from Mercury to Pluto, the evidence is all there.

Always crucial to the understanding of nature's planetary process, is the question of angles and planes in relation to that fundamental plane, the equatorial plane and pole of the Sun's particular motion, with which all our planets are destined to merge. Planetary motion in the Solar system shows that all of the above-mentioned angles and planes are undergoing change; axis and equatorial planes relate to spin and the orbital planes related to linear motion are all subject to change.

It is significant that all planets from Earth outward from the Sun have moons, several in considerable quantity. This suggests that moons are absorbed through time as the planets approach closer on their journey to the Sun. The

Chapter V – The Solar System

planets themselves therefore are transformed with each absorption; a transformation into a greater sphere. The two inner planets having long since absorbed theirs, whilst Earth is left with one yet remaining in orbit, it is very likely that there has been a number of other moons of various sizes and orbits previously orbiting Earth and eventually merged with our planet, this likelihood opens a new field of enquiry. When focussed on such a possibility, geological research could possibly ascertain the existence of Earth's previous moons and even explain some hitherto mysterious phenomena such as the demise of dinosaurs and other species.

Of the outer planets, Uranus and Pluto both with regard to their particular motion are rotating in a retrograde direction. Uranus, with an obliquity to its orbital plane of 97°.77. has a mere 8 deg. of correction to progress, as influenced principally by the particular motion of its primary the Sun. It will then become prograde in rotation to equate with the prograde direction of its orbital plane. Unlike that slowly retrograde rotating inner planet Venus, Uranus is rapidly rotating and has plenty of time on its journey to equate its equatorial plane with its orbital plane. Pluto the outermost planet is also rotating in retrograde fashion with an obliquity to its orbit of 122°.53 requiring correction through 33° to bring it into prograde rotation. It is seems quite natural to find that two of the outer planets are still undergoing the transition from retrograde rotation to prograde rotation. The other outer planets, Mars, Jupiter, Saturn, Uranus and Neptune are all rotating in prograde fashion.

On Earth, it should be expected that a movement of the pole and plane of the ecliptic is to be found as our planet accommodates to the gravitational demands of the Sun to draw its orbit into line with its equatorial plane; just as the Moon has to accommodate to the similar gravitational demands of the Earth drawing the moon to orbit in the plane of the Earth's equator. There is no angle or motion fixed and absolute in planetary motion, all are in a process of change. In the study of the conditions that provided and still support the continuance of life on Earth we have to take account of the changes in angles of planes and poles.

Similarly the spin of our planet is seen to be a movement subject to change, it is slowing, and the day is getting longer. This tells us that Earth was once spinning much faster and it is logical to predict that the Earth will eventually lose its particular motion completely, probably long before merging with the Sun. By comparison slowly rotating Mercury, takes 58.6 (Earth) days to complete one rotation about its axis, with an orbital period of 88 (Earth)days.

A planets rotation period will be slowing its days getting longer while, on the other hand, Orbit periods will naturally be reducing, consistent with, the closer the planets orbit gets to the Sun the less distance is the orbit the more rapid its orbits appear. This produces shorter orbital periods giving the impression of acceleration of a Planets orbit. There is an accelerating factor due to gravity that must come into the orbit equation too. But the shorter orbit distance must be a factor affecting the appearance of acceleration.

Even slower, Venus takes 243 Earth days to complete one rotation about its axis, almost equal to its orbital period of 224 Earth days.

It is said that the Earth's slowing rate of rotation is due to the Moon's tidal interaction but it is more in keeping with the laws of planetary-motion that the Earth's loss of particular motion is mainly due to solar tidal interaction. It follows that the Earth Moon tidal interaction, also in keeping with the laws of planetary-motion, was responsible for a slowing rate of rotation of the Moon continuing until its rotation ceased altogether. This is in keeping with other planets in our system, where similarly satellites lose their particular motion to their primaries tidal interaction. The fact that slowly rotating Mercury and Venus with no moons to supposedly slow their rate of rotation would indicate this to be so, it is more likely that in their case too, that it is the tidal action of the Sun that would be slowing their rate of spin.

The positive and negative relationship between satellite and primary is absolute, the primary being dominant throughout. The primary's whole function is a process of negation of its satellites. First is the negation of a satellite's independent motion about the Sun, next is the negation of its particular motion and lastly the negation of its independent existence as a sphere. This is the process of accretion of the spheres.

The Satellites

While it is quite accidental that a number of the outer satellites, of the outer planets, have entered into various orbital planes all in retrograde direction, however it is not accidental that the orbital planes of the inner satellites of the outer planets, are prograde and more in alignment with the equatorial plane of their primary than are the outer satellites being of later capture. And we may consider that our Moon has been and possibly still is, similarly affected by its primary Earth, in that, with each orbit, it too, gravitates incrementally toward Earth's equatorial plane. It is also significant that there is an absence of axial rotation of many of the innermost satellites of the outer planets; these are recorded as being in "synchronous rotation with their orbital period". This is better explained by recognising that satellites gradually lose their particular motion as they approach their primary until they lose their rotation altogether eventually to remain polarised gravitationally, to their primary while continuing to orbit closer to their primary

The polarisation of the moon, is nothing more than the relatively stationary gravitational bulge on the Moon since it is no longer rotating on its axis. Whether rotating or not, it does not alter the gravitational relationship. Saying that a satellite is in synchronous rotation is a fallacy since only for a moment of its orbital life will a satellite pass through, (as it loses its rotational momentum), a moment of semblance to synchronous rotation. The logical indications are as indicated by Venus, which has passed that moment (of rotating once in 243 Earth days in sync with an orbital period of 224 Earth days) where it resembled a synchronous rotation and continues to lose its rotational velocity, slowing to its present 243 day rotation.

All significant satellites as they approach their primaries must pass through Roche's limit (about 2.45 times primary radius) which effect on a satellite has been compared to Saturn's rings; whereupon tidal effects tear apart the satellite in what is surely the first stage of the final accretion process.

Our Moon's particular motion, (its rotation about its axis) has long since slowed to the point where it has ceased to rotate on its axis altogether, having lost its particular motion, yet of course its general motion is still much in evidence. There is no escape for the Moon trapped in the accretion process, though its distance from Earth is said to be still many times Roche's limit.

Just as Earth's Moon has lost its particular motion to its primary, similarly there are many moons elsewhere in the solar system that have lost their particular motion. All of these are orbiting about the equatorial plane of their primary. Astronomers also deny the absence of rotation of these many moons. Though, of little consequence since their tidal bulge is static, they prefer to claim that they "rotate on their axis synchronously with their orbit period". When properly considered, such a condition is 'perpetual motion' since any change in the rotation of Earth, such as slowing, or of moon moving closer or further away immediately upsets their synchronous theory.

It has been estimated that the effect of the terrestrial tide on the Moon, is 25 times stronger than that of the lunar tide affecting Earth, so the slowing effect and internal stresses on the Moon when it was rotating on its axis must have been enormous. The Earth, obeying this law as a satellite, will be losing its rotation to its primary the Sun.

In accord with the laws of planetary motion, the Moon is moving in its general motion closer to the Earth, just as all satellites move toward their primary; as much as many astronomers would deny this and have it moving away. Many astronomers would have us believe that satellites can be moving away from their primaries. Current astronomy asserts that the Moon is "accelerating" thus supposedly reversing its orbital movement closer to Earth by invoking that of an outward spiral, (a generate motion) moving Moon away from the Earth. However, a perfectly sound explanation for this apparent phenomena is given by a 19th century astronomer and geometrician (Col. Drayson), revealing that this apparent "acceleration" is accounted for by the annual precession of the polar axis of the Earth, the

Earth's second rotation; creating the appearance that the moon arrives at a solar eclipse earlier, or in a shorter period of time than at the previous eclipse, thus construed as an actual acceleration of the Moon in its orbit.

It is for this very important reason, to draw attention to Col. Drayson's fine exposition of this subject. I feel compelled to reproduce in the near future an 'Explanation of his work 'The Apparent Acceleration of the Moon's Mean Motion' and also his work 'Cause of the Supposed Proper Motion of the Fixed Stars' both based on Earth's second rotation.

The Accretion Process

Astronomy in general, it appears, does not openly acknowledge that all the matter in the Solar system is gravitating to the Sun, directly in the case of the planets and indirectly in the case of their satellites. For example, the notion is generally supported, particularly in the study of the motion of some of the satellites (moons), that their motion is suspended "tidal locked". There is no intimation given that these satellites tidally locked or not, must still obey the same gravitational law of planetary motion that their primaries obey in orbiting, as satellites closer to their primaries. On the contrary, the notion is conveyed that, not only are they not gravitating to, or closing on their primaries but can even be moving away from their primaries. Such a notion is an example of the mechanical materialist views dominating the science of astronomy. They don't realise that even if a satellite was movong away it could only do so by moving spirally.

In considering the Solar system, a system of spheres, from the dialectical-materialist point of view, the first observation to make is that nature produces no perfect circle in planetary motion – such would be perpetual motion. The law of gravity operating in the Solar system is absolute, drawing all satellites by means of spiral orbital paths to their primaries and all planets to the Sun. From the above arises the dialectical-materialist theory of planetary motion in contrast to the current mechanical-materialist viewpoint that ignores, or pays little heed to the above criteria when it suits their metaphysical arguments. No greater example of this is demonstrated in the assertion that despite the attraction/gravitation between two bodies in the Solar system, a particular satellite is moving away from its primary, being repelled from each other. The two bodies referred to are none other than the Earth and the Moon. The need to make such an assertion is firmly founded in and serves the subjective motives of the established scientific circles of astronomy and politics, also it serves established science to regard another planetary movement in a mechanical way by asserting that precession is a "wavering" or "oscillating" closed circular one, a strange conclusion indeed of observed planetary motion.

As each satellite of the Sun spirally orbits inward closing on the Sun, so too, each orbit progressively advances the plane of its orbit to eventually equate with the equatorial plane of the Sun. This reveals the dual function performed by the spiral in planetary motion. Here then is displayed in all its simplicity, a law of planetary motion that governs the movement of all our planets as they progress towards their elimination into the Sun.

In Fig,2.1 It can readily be seen the orderly process of accretion in the final stages of development of the planets in the Solar system. The three inner planets reveal clearly the order of approach to the Sun.

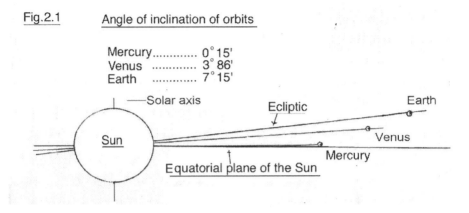

Fig.2.1   Angle of inclination of orbits

Mercury............ 0° 15'
Venus   .... 3° 86'
Earth   .... 7° 15'

When closely investigated it will be understood that the only way that an orbiting body can accomplish the movement to change its angle of inclination, to align with the equatorial plane of its primary is, through the deflection induced by the Sun creating the spiral path of the orbiting satellite. The gravitational force and the direction of rotation of the Sun create the dual action of the spiral orbits of the planets. The first aspect of this dual function is of course is the gravitational attraction pulling the Earth off its linear course, to orbit spirally closer to the Sun.

The second aspect is a response to the direction of rotation of the primary body the Sun. This aspect of spiral motion I refer to as 'progression' whereby the plane of the ecliptic moves, spirally in prograde fashion advancing the equatorial plane to equate to the equatorial plane of the Sun. This prograde inclination movement of 0".1247 has long been recorded annually along the equatorial plane and erroneously defined as 'Planetary Precession' a 'constant of precession contradicting that all planetary motion is produced by spiral motion, there is no escape from this fact.

The angles and planes of Earth in its travel about the Sun are shown, all of which are undergoing change, affected

Chapter V – The Solar System

by the dual nature of the spiral in planetary motion. Depicted in **FIG.2.2** is a view of the Earth in relation to the equatorial plane and pole of the Sun, about which all the planets of the Solar system are compelled to orbit. In this work on planetary motion, the angles of the planes of all the planets are similarly considered and related to the equatorial plane of the Sun, not on the plane of the Ecliptic as is customary in viewing the Solar system.

The close similarity of the outer planets inclination to the Sun's equatorial plane, compared to Earth's 7°.15:

Angle of Ecliptic to Equatorial plane of the Sun 7°15

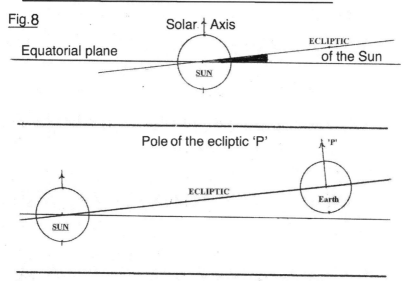

Angle of North Pole 'N' to the pole of the ecliptic 'P' - 23°28' 0"

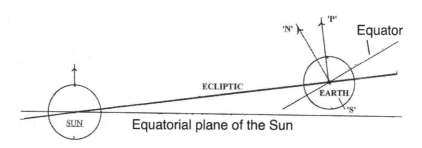

Alan Edward Dover | 45

Mars   5°. 65
Jupiter 6°. 09
Saturn   5°. 51
Uranus   6°. 48
Neptune 6°. 43

All the angles in planetary motion are undergoing secular change as the planets move in good order, spiralling in toward the Sun. The recorded annual changes reveal the laws in force begging for recognition in planetary motion demonstrating the laws by which the planets are governed to travel on their journey to the Sun in good order'.

CHAPTER VI

A GENERAL VIEW OF EARTH ORBITING THE SUN

Transition of Planets From Retrograde to Prograde Orbit

We next explore the ability of a rotating sphere to correct a satellite's retrograde orbit to prograde which rests with the susceptibility of the orbiting body to be gravitationally induced to spirally progress over the poles of the primary. Retrograde orbits in this way., are able to play their part in the actual accretion process of satellites with their primaries, all which eventually take place in prograde orbit.

The control and if necessary reversing of the direction of the orbit of a satellite about its primary, is a simple and gradual functional capability of planetary motion, since it is obviously an objective dialectical process in our Solar system of satellites to orbit and eventually merge with their primaries, upon reaching an orbit about the equatorial plane of that primary. It is logical to expect that there is a natural law governing the whole gravitational sphere of influence of a primary, into which, a satellite enters and which will s unifying process about in the most compromising and natural way. As the science stands today, resting on Newton's law of gravitation and Kepler's laws of planetary motion, no further advances have been made toward a greater understanding of planetary motion, in particular, the field of spiral motion and its dual nature producing orbital progression.

While science especially in the second half of the 20th century neglected the natural geometric aspect of planetary motion they are still firmly committed to the mechanical interpretation of those laws of planetary motion in spite of their space technology, computation of rocket trajectories and orbits of artificial satellites, tracking data etc. For this, mathematics and the definition of the law of gravitation serve their purpose. Yet so much remains unexplained in this natural science regarding planetary - motion. It is wrong to consider that all is explained by Newtons and Kepler's laws, they are not conclusive in explaining the dialectical, evolutionary and revolutionary transformation process of development of the planets and the Solar system. The hypotheses on the subject of the transformation of retrograde orbits to regular orbits through a process of orbital progression, brings to bear a case in point.

It is generally acknowledged that the orbital direction of satellites about their primary corresponds to the direction of the axial rotation of the primary, in particular the inner satellites; yet because of their fixed notions, it is not generally accepted that the retrograde satellites are obeying the same laws as the other planets and satellites, in spite of their apparent errant motion. Confusion will continue until the spiral mode of planetary motion is understood.

All orbital motion is not singular but has a dual function; satellites also, as they draw closer in their orbital spiral also advance their orbital plane (orbital progression). This is indicated by the planets in their orbits about the Sun, all, as they draw closer, at the same time seek to equate their orbital plane with that of the Sun's equatorial plane. There is no reason to suppose that the planets initially held and always maintained their present orbital planes in relation to the Sun. They are, as can be determined, gradually and continually adjusting to accommodate to the equatorial plane of the Sun. From Mercury with only 15'0" of inclination to that plane, through to that of Pluto with 24° 15' we can see the obvious established tendency. We have no planet in our system whose orbit is inclined greater than Pluto's 24 degrees to approach that of 90 degrees and beyond, whereupon orbits about the Sun would be retrograde. Yet there is reason enough to suppose that some of our planets could well have been retrograde in their distant past, such a conclusion we can draw from an examination of the moons or satellites of our planets which obey the same laws of planetary motion, where we find a number of them, significantly the outer ones, displaying the much steeper angular orbits indicating the greater inclinations upon which some initially entered their orbits about the sun. Thus orbital progression accommodates the satellites that enter into retrograde orbits about a primary, by transforming their orbits through the necessary progression over the primary's pole. It is therefore suggested that in the motion of the spheres there is no barrier to the transition from retrograde to prograde orbital motion, the change is smooth natural and of course relative.

We must bear in mind that there are two processes that take place in the solar system with regard to the adjustments of a planets orientation to the Sun; one is the process whereby a planets axial and equatorial inclination changes to comply with that of the ecliptic axis and the other is the adjustment to a planet's orbital inclination to comply with the Sun's equatorial axis and plane. This arises as the inherent dual interdependent aspects of planetary motion namely the general and particular or orbit and spin.

Linear motion or orbital progression proceeds from the moment that a planet comes into its primitive existence in the primordial mass, orbiting under the Sun's influence. At the same time, in the formation process the proto planet acquires spin or axial rotation, they are two separate but interdependent movements of a sphere acquired at the beginning of their formation. Neither of their separate orientations are fixed in relation to the Sun's Axis and equatorial plane. but due to the wonder of spiral motion the Sun can be seen to be bringing all planets (with one apparent exception, Venus.) to comply with a law that the planets adjust their final orbital orientation to approach the Sun, orbiting on the Sun's equatorial plane and their axis and rotation to comply with that of the Sun's own axis and rotation. All is accomplished with spiral planetary motion.

All our planets have each grown through this same process of accretion. No doubt much of the process of accretion has been the result of collisions by small debris especially in earlier stages of our systems development as we witness the numerous craters abounding even on our own planet but Generally speaking planets do not collide. Witness all the moons in the system, lesser planets that have all the features of the greater ones. There is plenty of evidence of asteroid like materials colliding with the planets yet planetary motion itself appears to obviate collisions as such by the more gentle accretion process. We have to acknowledge the strength of the evidence of the spiral as a law of planetary motion. The formation of the sphere in itself as a basic form of matter throughout nature in all magnitudes is a wonder in itself.

Observation indicates, as Earth orbits the Sun each year it moves ever closer to alignment with first the plane of the ecliptic and eventually the Solar equator. As we look at each of our sister planets we can see them all in different stages of their adjustment in this process of orientation to the Sun.

Explanation to Fig. 5.1

Let 'orbital progression' be defined as the annual advance of the orbital plane of a satellite toward alignment with the equatorial plane of its primary by means of the rotation of the Line of the Apsides.

The circle presented in **Fig. 5.1** contains quadrants around the Sun oriented on the solar axis and its equatorial plane of rotation.

The spiral mode of Earth's ecliptic cycles as it spirals closer to the sun and advances, changing obliquity through the quadrants, reaching Zero obliquity at 'B'

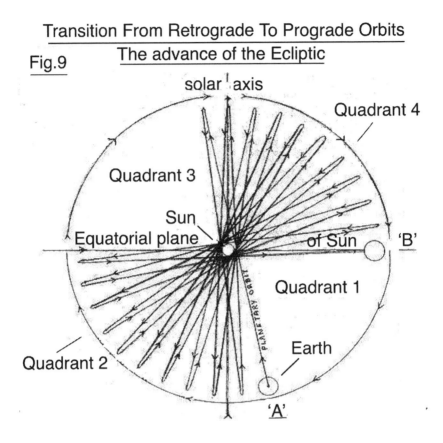

Transition From Retrograde To Prograde Orbits
The advance of the Ecliptic
Fig.9

Here orbital motion about the Sun is traced in exaggerated increments illustrating the likely transition process from retrograde to prograde orbits. Progression ceases when the planets orbital plane reaches alignment with the equatorial plane of the Sun. This latter is indicated by the fact that the obliquity of the ecliptic is annually decreasing at an increasing rate, therefore slowing as the Earth's pole approaches the pole of the ecliptic.

To explain the capabilities of planetary motion to convert a retrograde orbit of a satellite about its primary into a prograde orbit, we merely need to see the effect of a possible movement of a planet through orbital progression as follows:

The range that a satellite can change to prograde orbit about a primary is 90 deg. above or 90 deg. below the equatorial plane of the primary and will only require a correction through progression, of a maximum of 90 deg. above or below the equatorial plane to bring a planet or satellite's orbital motion to conform to a prograde orbit.

In short, wherever a planet or satellite enters into retrograde orbit it stands to be corrected and brought to prograde orbit in readiness for assimilation and union with its primary.

It can be said that the attractive forces of the primary body (generated and governed by its own particular movement, its direction of rotation) will influence any orbiting body with a retrograde orbit, to conform to its own direction of rotation, a law apparent in the dynamics of planetary motion.

Explanation to Fig. 5.2

Fig. 5.2 offers another view to understand the two angular changes taking place at the same time in planetary motion, here can be seen the orbital progression cycle of the planet Earth together with its precession cycle in pure and simple format. The view illustrates:

1. Precession of the pole 'N' about the pole 'P' is bringing the Earth's equatorial plane to align with the plane of the ecliptic.

2. Progression of Earth's orbit from 'E' to 'F' is bringing the Earth's orbital plane to equate with the equatorial plane of the Ecliptic and eventually, to the equatorial plane of the Sun.

The Dynamics of Spiral Planetary Motion

The Earth is depicted in its present position of the orbital progression cycle at 'E'. The two inner planets Mercury and Venus orbit between 'E' and 'F'.

We can ascertain from the diagram that the Pole of the ecliptic would change its inclination to the solar axis in its passage from 'D' to 'F' on the progression cycle by 45°. Primarily the diagrams purpose is to focus on the slow change of the angle of the pole of the ecliptic to the Sun's equatorial plane and pole. The ratio that needs to be determined, is the amount of descent of Earth on its *progression* orbit cycling during the period of one 26,000 years *precession* cycle. In **Fig.5.2** arrows on the orbital perimeter indicate direction of orbital progression bringing Earth's orbit towards alignment with the equatorial plane of the Sun at 'F'.

Fig.10

Arrows on orbital perimiter indicate direction of orbital progression, bringing Earth's orbit into alignment with equatorial plane of the Sun at 'F'

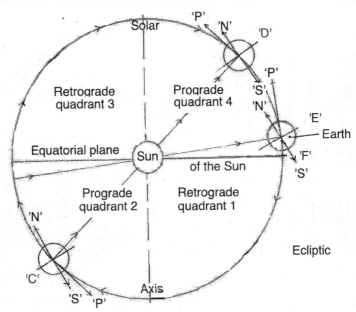

Chapter VI – A General View of Earth Orbiting the Sun

Note - It will be noticed that the diagram depicts N as always pointing to the same point of the celestial sphere throughout the progression cycle. Whereas **N**, in fact would be continually cycling about '**P**' (the pole of the ecliptic) during the progression cycle and would be previously cycling, on a much wider angle of obliquity than the present 23.5° of obliquity.

The diagram is interesting too if we consider the Earth in the distant past in the higher region of Quadrant 4. The Earth's pole of daily rotation **N** which has always been under Luni-solar influence The Pole **N** could have been inclined 40 to 50 deg. of obliquity or more, to give explanation to such evidence as the previous existence of tropics in the polar regions. This is opening up a new approach to fields of scientific investigation in Astronomy and Geology.

In **Fig.5.1** we followed the orbital path of Earth as it progresses clockwise over the Sun's north polar axis into quadrant 4 becoming prograde in orbit, moving on in **Fig.5.2** through '**D**' toward '**E**', This movement is caused by the sun's attraction inducing the spiral movement of the earth's orbit to progressively align with the equatorial plane of the Sun by means of the spiral rotation of the Line of apsides.

At the same time as the Sun is affecting the Earth's orbit, it is also affecting the Earth's spin by inducing it, as Newton indicated, to align its equatorial plane with the plane of the ecliptic. This is accomplished as explained, through the spiral rotation of the precession cycle.

We can now visualise from the Earth in its flight about the Sun, closing slowly with the Sun's equatorial plane from '**D**' to '**E**' while spinning but with its axis cycling about the pole of the ecliptic and rising, closing its inclination to the pole of the ecliptic. Its axis is cycling relatively rapidly in the opposite direction to its spin. Thus, we have a picture of two distinct aspects of planetary motion taking place simultaneously between a star and a planet or a planet and its satellite.

We could consider how many times that **N** 'cycles about the pole of the ecliptic '**P**' during the passage of Earth say, from '**D**' to '**E**'. the change in the obliquity we can see, would be quite considerable indeed,

No matter to what extent the actual movement that our axis has passed over, one can understand historically the enormity of the effect on the seasonal and geological changes that have been taking place on our planet throughout eons past. Thus, with this understanding as the basis of investigation, much of the mysteries about climate and geological changes can be better explained.

The spiral in planetary motion is acting in both the general and the particular aspects of planetary motion, so it is necessary to examine them separately although the difference is but relative. All is accomplished by the spiral in planetary motion:

(a) In a planets particular motion, its rotation; the Luni-solar gravitational influence tends to equate the planets equatorial plane with the plane of the ecliptic causing precession of the poles.

b) In a planets general motion, which is its orbital path as a satellite The Sun's gravitational influence tends to equate through progression the planet's path in the same prograde direction as its, the Sun's, own direction of rotation; in order to bring Earth's orbital path (and pole of the ecliptic) to eventually coincide with the Sun's own equatorial plane and poles.

Chapter VI – A General View of Earth Orbiting the Sun

We can now consider the effect the changing inclination to the Sun that Earth, in its passage through quadrant 4 has had on Earth's climate. Any change in inclination affects the ice and water balance, the size of ice caps and sea levels alone is open to new interpretations. Such changes would affect the conditions for life on Earth. The growth and decline of the ice ages are the subject for my next enquiry into the effects of spiral planetary motion.

By using **Fig.5.2** we can follow the possible cause of the critical continuous change that has taken place due to the more extreme movement of Earth's axial inclination to both the poles and planes of the ecliptic and the Sun. For example, if we traverse the spiral motion of earth about the pole backward, increasing the obliquity of the ecliptic in millions of years previously on its journey. We would be retracing its orbital passage past the point where it passed over the solar axis, from quadrant 3 into quadrant 4. In quadrant 3 as we saw it would be orbiting in retrograde fashion. Dramatically, at this point of Earth's orbit passing over the poles of the solar axis the plane of the ecliptic would be inclined to the equatorial plane of the Sun by 90 °. At this point it would be difficult to estimate the angle of the obliquity of the ecliptic since the great difference in the annual (or century) rate of change between the two interdependent movements, precession and progression, So really it is sensible for science to focus a study to what may be calculable, such as the passage of Earth from 'D' to 'E'. Which involves a comparison of the change of angle of ε per century (47" °) with the change of angle of progression of the Ecliptic with the equatorial plane of the Sun per century. Today it is a mere 7° .15' and extremely slowly annually reducing.

We can then reasonably consider that at this point, the separate precession movement of Earth's pole, while decreasing the obliquity of the ecliptic, will have a very much greater angle of obliquity of the ecliptic as it cycles about the pole of the ecliptic, even at some stage, as great as 0° obliquity of the ecliptic. Under these conditions, at some stage 90 degrees obliquity the Sun would be passing annually over the Earth's Poles, the transition from sun light to darkness would be a bi-annual event, not at all suitable for life as we know it.

Of course this is speculation, the truth of the matter depends on calculations of the two decisive factors, they are first, the annual rate of progression of Earth on its path about the Sun equalising the ecliptic with the equatorial plane of the Sun and the relatively rapid cycles of the 25,000 year continuous precession of Earth about the pole of the ecliptic. To determine the ratio between the two is a challenge to the science of astronomy of the future, at least if we move to the Earth's past passage approaching 'E' It is quite feasible that examining a few seconds of arc backward from 'E' 'toward 'D' with the known 12.47" per century movement of the orbital progress of the Earth about the Sun and a similar sum of precession per century will give some idea of the more recent geological changes that have taken place. This is surely within the bounds of present day technology.

The Life Cycle of a Planet is Governed by Spiral Motion

Since there is no such thing as perpetual motion, all observed orbital motion in the solar system can only be a spiral, moving toward a centre. It seems that the gravitational influence affects the satellite in a two-fold manner; whilst creating the movement toward the primary body, it at the same time deflects or influences the path of the satellite sufficiently with each orbit, to reduce its inclination, to ultimately advance the orbit. And, if necessary (when retrograde) over a polar orbit then through 90 degrees to correspond with the equatorial plane of the primary sphere, eventually bringing the satellite to a most natural position, the equatorial plane of the Sun, the point of entry, to merge with the primary. It seems that at that point when a satellite's orbit reaches equality with the equatorial plane of the sun that it will have lost most if not all of its particular motion and will not be spinning on its axis before merging with the sun, a position that our Moon has advanced to in relation to Earth.

The assertion here, is that the orbital plane of a satellite, under the influence of its primary is no fixed configuration but is undergoing change whilst moving toward the primary, the plane of that movement is itself advancing to merge with its primary on the equatorial plane and in the same direction. These phenomena when studied will be recognised as being among the laws of planetary motion, since the fact that most of the planets and satellites orbit the same way is not at all accidental.

Though not yet acknowledged, the study of planetary motion shows that though the orbital direction of a sphere about its primary initially may be retrograde the possibility exists in planetary motion for a body in retrograde orbit to become prograde in orbit to its primary, the dynamics of this is explained in **Fig. 5.1** and **Fig. 5.2**

From the existing recorded facts in present day astronomy, when correctly interpreted by dispensing with preconceived mechanical notions; it can be reasonably asserted that:

a) The essence of planetary motion is a dual functioning spiral motion – a fundamental law governing planetary motion.

b) The difference between planet and satellite being relative, planets are but satellites, both are subject to the law of spiral orbits.

c) The fundamental spiral is evident throughout the life cycle of the sphere from its formation (accretion process) to its demise on joining as a satellite with another larger sphere and ultimately with the Sun.

d) The relationship of the sphere's direction of rotation with the direction of its orbit, arises with the particular rotation of matter within the sphere of the general primordial mass of matter.

In the formation of the new, possibly small sphere, it would continue in its general motion, orbiting about the centre of the forming Sun. In the course of its passage it will continue to attract lesser formed spheres to itself as satellites, absorbing them and becoming larger and may even itself, become a satellite attracted to another larger sphere and be absorbed in the protracted spiralling process of accretion There is abundant evidence of the massive stratified layers of rock about our planet building our sphere during the formative period of the accretion process. Even today, our solitary remaining satellite, the Moon, smiles as it bears down on us as living proof of the process.

CHAPTER VII

THE PLANETARY PRECONDITIONS FOR LIFE ON EARTH

> It is essentially planetary motion that has provided the unique and primary conditions for the spontaneous development of life forms leading to such as presently exist on Earth. The presence of our solitary Moon takes pride of place in this phenomenon, a key figure for us in the 'Solar ballet of planetary motion'.

Here on Earth, we humans and all higher forms of life are the product of a unique, temporary and relative phenomena of matter in motion, of spheres and their motion that have created optimum conditions that will only exist for a limited period of time to bring forth those life forms. Nature has momentarily created a relatively stable environment for the spontaneous development of bio-chemical action, to advance to a very high degree, the highest being, as Frederick Engels said, "to thinking", to the manipulation of concepts, to ideas, which is undeniably, the highest form of motion of matter, a faculty most advanced in the human brain.

The origins and foundation for development of life on Earth remains a great mystery until we clearly identify those optimum conditions that maintain the stability needed for further development out of those first, and most primitive biochemical life forms, dependent as they are on the daily, monthly and annual stimulus. Not least of which we note is the vital daily tidal action a continual, regular and uninterrupted repetitive movement over a very long period of time we owe to the singular presence of our solitary moon.

These unique and optimum conditions that exist here on Earth and are not yet found elsewhere within the bounds of our perception in the Solar system, much less beyond in the celestial magnitude. It is not at all an over simplification to state that in general and in the first instance, that life on Earth owes its existence to the simplest form of matter and its motion in nature, the sphere and its motion. To understand planetary motion is to understand how life as we know it was able to emerge.

Obvious as this might seem, yet full investigation of the motion of the spheres has not been conducted by science. This investigation is necessary to advance to a better understanding of the unique stable conditions necessary for the emergence of primitive bio-chemical forms from which the higher life forms arise. Well known are all the other factors such as the effect of the Sun, giving tidal effect, seasons, heat, and light on our planet. Then we have that vital element, water, that responds to the solar stimulus giving us a favourable climate engine, together with the actual material composition of Earth itself. Known too, is the importance of the daily rotation, Earth's motion, its polar attitude. Yet the conditions existing on Earth are never considered without its companion the Moon and those vital life giving Lunar tides, leaving only those Solar tides. Such an equation can probably only produce at best the lower forms of life such as mosses, so we understand we owe much to our solitary Moon.

In previous chapters we looked at those obvious aspects of the motion of our own spheres beginning with the relationship between the Sun and her planets and most importantly their orbital inclinations to the equatorial plane of the Sun. The fact that they are spheres compels them to move, as do all spheres in the Solar system in accord with the laws of planetary motion and as we discover, the laws of planetary-motion are not fully understood by science. That, in many instances, science has ossified what is known, into metaphysical lifeless obstacles to our further understanding, all to suit particular human ideological and political interests. In particular we find spiral orbits defined as eternal circles or ellipses and others seen as immutable retrograde orbits. Even great mathematical studies are made of sections of orbital motion, an infinite study of "curves" and after all, any curve is merely a part of a circle or a spiral. Our scientists unfortunately prefer to relate a curve to circles and never to treat a curve as part of a spiral orbit therefore, are at odds with nature. Together, their pursuits constitute severe limitations to progress in understanding those unique conditions for life on Earth and for this reason we are bound to investigate and further develop our understanding of the laws of planetary-motion

The Unique Conditions – The Earth Moon Union.

As we know, the general condition created by the Sun through its light and radiant energy is vital to the development life on Earth. Its gravitational influence adds to the Moon's relatively greater gravitational influence causing tidal effects. Other factors come into play and it is with these other factors we must concern ourselves too. Planetary motion shows us that we have to take account of the effect that the factors of proximity and mass of these three spheres have now and have had on each other in the past, since all are subject to continual change.

At this time, on its own the Sun could not produce life on Earth as we know it, since its gravitational effect would produce only a very passive internal movement in Earth. It is likely that only the lower forms of life would be developed and maintained on Earth by the Sun's tidal influence in the absence of our very important satellite, the Moon. Yet we may consider all things being relative, the enormous effect a given satellite such as the Moon can have on its primary the closer it gets, causing tremendous upheavals. Such abundance of evidence of what has been inflicted by Earth's previous satellites merging, is plain for all to see on the geological layers composing its surface. The effect of the actual proximity of the Moon at this point in the Earth-Moon development, produces a very favourable delicate tidal balance in its relationship with Earth. It will be realised that when our existing Moon was much further away on its approach, the tidal effect was much less to a point which Earth would receive little effective tidal movement to induce the development of life forms, perhaps little more than to promote the most primitive of life forms. We have to realise too that there may have been a number of other larger and smaller moons orbiting together that preceding our existing moon too. Such a condition even with two moons would create tidal chaos. However, now with one solitary moon we have the ideal tidal balance for life.

Of course, when the Moon comes much closer to Earth, the tidal effect will eventually become enormous to the point of destroying and prohibiting at least all higher forms of life through very violent tidal action. So the Moon's proximity to Earth is of great importance with regard to its contribution to the creation of the conditions for the development of life on Earth. Equally so is the size of our Moon is a factor, since if it were smaller or larger and of less or greater mass then the changes in the tidal effect on Earth could be similar to that of proximity. So as we can see the size and distance of our Moon, and of course Earth have been critical factors in contributing to the optimum conditions for the development of life on Earth.

Just as much too, is the Earth's proximity to the Sun which plays such an important role in the producing the delicate balance of our climatic engine, our weather for example. Planetary proximity then, is also critical in producing conditions for life on Earth. So it is of great value to science to understand the continual protracted changes taking place in the proximity of the Moon to Earth and the Earth to Sun, in fact a better understanding of the laws of planetary–motion.

Due to the difference in distance from Earth of the Sun and the Moon, their vastly different gravitational forces reach us proportionately different, the Moon having greater gravitational effect on the earth over that similar tidal effect of the Sun. The proximity of these three spheres to each other at this time is the first part of the explanation whilst the second part lies in the general and particular motion of Earth and Moon or orbital periods and rotational periods.

The Earth has its daily rotation. The Moon however, does not rotate on an axis at all, having lost its particular motion, it is at present a passive orbiting partner in the impending union of Earth and Moon. These general and particular motions of the spheres all combine to create the actual preconditions for life on Earth like the general (orbital) motion of the Earth combined with its axial inclination producing the cycle of regular annual seasonal variations. Then of course we observe that Earth obeys the law of planetary motion that governs its direction of rotation to coincide with the direction of rotation of the Sun.

The juxtapositions of these spheres, Sun Earth and Moon as will be readily understood, is temporary and relative; all subject to secular change. The conditions that they have created here on Earth are but a fleeting moment in the passage of time and motion within the Solar System.

From observing our sister planets and their satellites we may consider the likelihood that the three inner planets too, once had one or more satellites on their journey to the Sun like the outer planets. And of course we must ask, what conditions would exist on Earth under the gravitational effect of two or more Moons with possibly orbits inclined to each other. Again, size and proximity are important factors too. Surely we must conclude that such conditions would not be stable enough for even the most primitive life forms to be generated.

When we consider the development of life itself through the vital effect of the Moon's tidal friction on Earth we must speak of course of the corresponding effect of the daily rotation of the Earth upon its axis causing a precise and regular tidal rise and fall of the terrestrial mass twice daily corresponding to an unfailing heartbeat in a living organism. The strength of the beat varies only marginally,

Chapter VII – The Planetary Preconditions for Life on Earth

corresponding to the opposition or conjunction of the two spheres the Sun and Moon in relation to Earth..

What has been produced by this life giving heart beat is known to biochemistry as the emergence of primitive forms of life the single cell then with creatures appearing first in the seas being the progenitors of more complex forms, thus commencing a new and generate higher form of motion of matter, created by these real and unique material conditions. And let it be understood that the single most important factor in the equation is the very presence of our solitary Moon. The truth of this fact can be verified by the comparisons we find by studying the relationship and motions of our sister planets and their Moons all displaying the inherent laws of orbital motion in our Solar system.

Higher forms of life on Earth could only develop under the calmer regular conditions existing on Earth in the period, of one solitary Moon orbiting Earth, one solitary Moon being the deciding factor to complete the equation for the development of higher forms of life on Earth. So much depends on the regular diurnal tidal and periodic effects of our solitary Moon.

Contemporary astronomy and consequently geology and the like, completely miss the importance of the role of our Moon in creating the conditions for life on Earth, "Life" (meaning biological development to the higher forms). Whereas, under the conditions of our planet with no Moon in attendance, the Earth, having consumed the Moon, finally Earth will be transformed into a greater sphere, in the process, all existing life forms will be obliterated.

Under the influence of solar tides alone, life on Earth would most likely regenerate but be restricted to generating lower forms of life. And this, only on the assumption that the planet would once again, after cooling, have similar chemical composition with seas and a suitable atmosphere.

Our standpoint in this study must of course always reflect the standpoint of the motion of all the planets without exception and that is, that all motion in the Solar system is moving and related primarily to, the equatorial plane of the Sun. The laws of planetary motion dictate this to us.

Geology –Stratification of Earth

The science of Geology too, is not blameless but is stifled as a consequence of the 'invariable' solar system notion, inspired by astronomers and the establishment,. Also we ask Geologists, why has the mass of evidence, the sheer immensity of layer upon layer of stratification not awakened geologists to the fallacy of oceanic and flooding 'sedimentation' as the origin of all strata? It is quite ludicrous to say that all these layers upon layers forming the earth's crust originate from a process of sedimentation alone.

It doesn't seem to occur to geologists that this mass of matter could have periodically descended down through our atmosphere from various satellites such as moons over many millennia. On inspection the layers are clearly interspersed and even mixed with local volcanic ash and debris and of course with some evidence of oceanic flooding and sedimentation areas too. Sedimentation however cannot account for the huge amount and depth of the layers forming Earth's crust.

The whole process of planetary motion is one of accretion. When considering the subject of accretion in the Solar system, we refer to the current view of astronomy which considers the formation of the planets from the earliest stage of accretion in the Solar Nebular of matter, "colliding" into primitive bodies "clumps" to attract matter, becoming proto-planets orbiting the Sun. Astronomy generally, does not view accretion as a process of development of the Solar System from beginning to end but only as a process in that early stage to the formation of 'clumps'. Also is defined the "accretion disk" formed about a celestial body. The many notions of gravitational force, of attraction and growth, are vague descriptions of later development. The definition of planetary accretion does not include planetary orbital motion but rather only the sticking together of particles through "collisions".

Accretion in the present stage of formation of the Solar system is not a process wholly of collisions, although even today many objects, not being planetary spheres are in solar orbit in random trajectories not orbiting any particular planet. These asteroids could still be responsible for 'colliding' with spheres, very much so in the distant past but today still likely even today even if rare. However they are subject to the accretion process orbiting the Sun if not the planets.

Viewing the planets of the solar system today; all in a very advanced stage of accretion, we see that the capture of smaller spherical bodies is a normal process. Gravity, the universal law of our system operates within the system from beginning to end. All the spherical bodies, satellites and primaries were formed through the earliest of stages of accretion in their individual orbits about the sun. Today, the later process of accretion of the spheres is somewhat different to that of the earlier process of accretion of matter, being a much more settled, orderly scene. With much fewer collisions of random objects striking the spheres. Now, spiral orbital motion is essentially the continuation of the process of accretion in the motion of the spheres. Accretion then is none other than manifestation of the law of gravity.

From the earliest stage when most if not all the particles of matter are condensed into numerous spherical bodies of varying size. Then with process of accretion continuing, due to the law of gravitation there begins the next stage. whereby, the smaller spheres, over great periods of time, will be gravitating to and becoming 'captured' by the larger or denser spheres. The general scene, is then one of many larger and or denser spheres each with numerous satellites contained within their individual field of gravitational influence, spiralling ever closer to merge with their primary and contribute to the growth of that primary in the accretion process. This developing scene would ultimately produce such as we could recognise as our solar system, a process which is continuing to this day.

It appears some satellites consumed in this process of merging, have deposited numerous even layers of strata anything from a few millimetres to many metres thick, each orbital layer of rock clearly distinguishable from the one before. It is likely that before and possibly during the satellite dissolving in the friction with Earth's atmosphere, the gravitational forces would be enormous. The Earth surface would be undergoing tortuous tidal upheavals gradually renting and twisting the previous strata.

It appears that the planets on their journey, consume all their satellites before reaching the Sun. Mercury and Venus consumed any they may have had while Earth, has now only one remaining of her satellites, while some outer planets have many satellites.

Chapter VII – The Planetary Preconditions for Life on Earth

All things in nature have a beginning and an end, science has generally understood the beginnings of our solar system yet the process of development of our system, its mode of development, planetary motion is not fully understood because it has been ignored. The reason for this is Religious, and political, so stagnation reigns. But regardless of man's ideas the process of development of the solar system continues, each satellite is bound to its planet and will eventually merge with it, just as all our planets, as satellites themselves must merge with the Sun. From the point of view of nature, it is a relatively gentle, orderly and wonderful process. The solar system is performing 'A ballet of planetary motion' that mankind at present, cannot see through the abounding mist of fixed notions concerning planetary motion.

CHAPTER VIII

A WEB OF DECEIT — THE 'INVARIABLE' SOLAR SYSTEM

> *Oh, what a tangled web we weave,*
> *when we first practice to deceive.*
> Sir Walter Scott

The Newton — Hooke Correspondence

The advance to date has merely raised the perception of planetary motion from a circle to that of an ellipse. In truth the orbital motion of a planet is neither, for such a closed orbit as a circle or ellipse is not a true scientific description of planetary motion. It is however, a suitable mechanistic definition for the epoch of industrialization but left unqualified, such a description is a portrayal of perpetual motion. It was not until 2013 that I had examined the Newton Hooke Correspondence and learned of Dr. R. Hooke's genuine approach to planetary orbits and his vision of the spiral orbit. The historic and noble quest for a better understanding of planetary motion had not been completed and since Kepler's time, been ignored or abandoned until Robert Hooke's initiative brought it into the light in the Newton, Hooke correspondence of 1679.

Chapter VIII — A Web of Deceit — The 'Invariable' Solar System

The question of orbital motion being an open or closed curve began to come into focus in the year 1679, when several notable scientists were considering the issue but it really came into the open as an issue to be resolved, as a result of the studies of Robert Hooke, who, as Secretary and curator of the Royal Society he approached Isaac Newton in correspondence invited him 'as a great favour' 24 Nov. 1679 to review his own contribution to the subject of the attractive motion towards the central body:

> *'if you shall please to communicate your objections against my Hypothesis or opinion of mine particularly if you will let me know your thoughts of that compounding of celestiall motions of the planets of a direct motion by the tangent & an attractive motion towards the centrall body'*
> *(R. Hooke 1679: correspondence)*

The development of the science of astronomy stood on the threshold of what should have been the greatest advance since Kepler. As the principal scientists of the day Newton and Locke had the opportunity to take the next great step for astronomy and celestial mechanics and geology. Hooke had taken the initiative in his 1674 work 'An Attempt to Prove the Motion of the Earth from Observations', in the Summary of which, he outlined his 'three suppositions':

> *'And shall only for the present hint that I have in some of my foregoing observations discovered some new Motions even in the Earth it self, which perhaps were not dreamt of before, which I shall hereafter more at large describe, when further tryals have more fully confirmed and compleated these beginings.*

At which time also I shall explain a System of the World differing in many particulars from any yet known, answering in all things to the common Rules of Mechanical Motions: This depends upon three Suppositions.

First, That all Coelestial Bodies whatsoever, have an attraction or gravitating power towards their own Centers, whereby they attract not only their own parts, and keep them from flying from them, as we may observe the Earth to do, but that they do also attract all the other Coelestial Bodies that are within the sphere of their activity; and consequently that not only the Sun and Moon have an influence upon the body and motion of the Earth, and the Earth upon them, but that also and by their attractive powers, have a considerable influence upon its motion as in the same manner the corresponding attractive power of the Earth hath a considerable influence upon every one of their motions also.

The second supposition is this, That all bodies whatsoever that are put into a direct and simple motion, will so continue to move forward in a streight line, till they are by some other effectual powers deflected and bent into a Motion, describing a Circle, Ellipsis, or some other more compounded Curve Line.

The third supposition is, That these attractive powers are so much the more powerful in operating, by how much the nearer the body wrought upon is to their own Centers. Now what these several degrees are I have not yet experimentally verified;

Chapter VIII – A Web of Deceit — The 'Invariable' Solar System

but it is a notion, which if fully prosecuted as it ought to be, will mightily assist the Astronomer to reduce all the Coelestial Motions to a certain rule, which I doubt will never be done true without it. He that understands the nature of the Circular Pendulum and Circular Motion, will easily understand the whole ground of this Principle, and will know where to find direction in Nature for the true stating thereof.' 'This I only hint at present to such as have ability and opportunity of prosecuting this Inquiry, and are not wanting of Industry for observing and calculating, wishing heartily such may be found, having my self many other things in hand which I would first compleat, and therefore cannot so well attent it. But this I durst promise the Undertaker, that he will find all the great Motions of the World to be influenced by this Principle, and that the true understanding thereof will be the true perfection of Astronomy.'

(Dr. R. Hooke 1674 'An Attempt to Prove the Motion of the Earth From Observations' pp.27–28)

When his "Ellipsoidal" sketch of (1679) is applied to his earlier three Suppositions in the Summary (1674), it can be seen that Hooke was on the verge of realising and proving the real explanation of the orbital 'Motion of the Earth'. He had brought the vital elements together to truly define the nature of planetary motion. We can therefore well understand his closing statements in the 'Summary'.

In the correspondence of the 28th of Nov. 1679, Newton replied that he had lost interest in natural philosophy and had no previous knowledge of Hooke's work on the subject.

The fact of the matter was, that Hooke was describing in his work, that orbital motion was a curved path gravitating to a centre. Clearly the description is nothing else but a spiral. The truth of which was at last forcing itself on scientific minds. Hooke's work shows that he was pursuing the study of orbital motion in the clearest and most objective way. His invitation to Newton to join in the topic, revealed the objectivity of his intent as curator of the Royal Astronomical society for a collective review of the subject. Newton of course seeing the importance of Hooke's intent now became, very interested in the topic of 'natural philosophy' and as a consequence, was later in 1684, to present his own version 'On Motion' (De MOTU) to the Royal Society, after Hooke's death.

Newton was to claim that he had previously come to Hooke's conclusion before him. If this were the case, the question must be asked, why he had earlier said he had 'lost interest' in natural science. However this led to the disassociation of the two in 1680.

After Hooke's death, Newton, with his prestigious reputation secure, chose not to follow through on the emerging theme of orbits curving to a centre and continued with complex mathematical discourses on curvature. These studies would dictate the course of research, thoroughly influencing the establishment of the day. The studies have ensued ever since and reduced the investigation to the mathematical analysis of curvature; which is accepted as the study of planetary motion. This situation has not led science to the obvious conclusion of the curve in orbital motion being part of a Spiral not part of a circle.

Chapter VIII – A Web of Deceit — The 'Invariable' Solar System

All interest in Hooke's initiative that would lead to the inevitable clear conclusion that orbital motion is a spiral was lost due to Newton; this, in spite of the fact that apparently in Newton's Dec. 13, 1679 letter to Hooke, he quoted a mathematical formula that indicated a case 'of orbital motion rotating toward a centre by an infinite number of spiral revolutions,' This brings Newton's motives under suspicion because he never mentions the word spiral again, so any advance was checked. We must consider, If Newton's subsequent actions were deliberate to suppress the notions of Hooke? However, in the correspondence, Hook had produced a sketch of his vision of a spiral orbit and so was the first person to describe spiral orbital motion, naming it the 'Elliptispiral'. He had a superior vision of the motion, compared to Newton's sketches on the subject of 'a falling body.to the centre of the Earth'. Had Locke's spiral image of orbital motion been adopted and linked to Kepler's elliptical orbit the subject of planetary motion would have been liberated from the stagnation that has existed ever since.

It shows that by 1679 Hooke had begun to recognise and roughly depict that first 'principle' that influences all the great, 'Motions of the World' the spiral, the common factor that served to explain his three suppositions concerning orbital motion.

It was a promising advance only to be stalled by Newton in the interests of the preferred notion of the 'invariable' solar system. However, it gives me great pride and satisfaction to able to contribute to the recognition of Dr. Robert Hooke's veracity and achievements in the quest for the truth concerning planetary motion.

The Dynamics of Spiral Planetary Motion

Although the social and political reasons motivating this corruption may well be obvious to some scientific observers, yet no one, apart from Dr. Robert Hooke in 1679 has seen fit to attempt to expose it. However, here, we are going to look at how this trend came to be instituted, for instituted indeed it is. The historical origins of this trend has established what passes for the dynamics of planetary motion even to this day; though more recently, opinion has forced the science to move away from the 'invariable' description to declare the order of the solar system as 'chaotic But nothing has really changed; the perpetually oscillating theories are still in place; and it must be said, that the Solar System is certainly not "chaotic", far from it, rather it is the scientific notions that are chaotic.

Currently however, there is growing interest in rehabilitating Hooke, to recognise him as the progressive scientist that he was, although, pathetically, much of the interest is merely to illuminate the question of whether

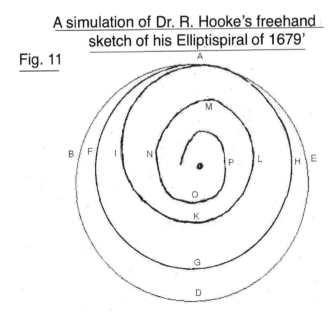

A simulation of Dr. R. Hooke's freehand sketch of his Ellipsispiral of 1679'

Fig. 11

Newton was pursuing the topic before Hooke or not. Was it Hooke who motivated Newton to produce his revered work 'DE Moto ' and Principia or not? The interest is trite compared to the advance that could have taken place; the real issue was and is, the fact that the Spiral, as the actual mode of planetary motion was forcing itself on man's consciousness; albeit to be delayed a little longer by Newton's subjective role. The Hooke – Newton correspondence of 1679–1680 did focus attention on Hookes work and his objective role awakening the further understanding of spiral orbital motion, to which he played an important part in the history and study of planetary motion.

The Distortions of Newton, Laplace and Newcomb.

What the science of astronomy interprets and accepts as the motion of our planet Earth, specifically with regard to the cause and effect, of the dynamics of its spin and orbit, contains gross distortions of the laws of planetary motion. It is clear that the reason for this corruption of the science, stems from an innate desire to present the Solar system as an 'invariable system'. As a consequence, the motions of the planets and their satellites have been presented by the champions of astronomy as systemic fluctuations, upon what are considered as 'invariable' axes and planes of the planets of the whole solar system, a view of a planetary system that will 'only oscillate about a mean state from which it will deviate but by a very small quantity' as Laplace was to put it.

This corrupted interpretation being presented as planetary motion is more akin to perpetual motion and surreptitiously applied to our solar system. Evidence will be found throughout our science from Newton through Laplace to Newcomb, upon whose calculations much scientific research has been based, through to the more recent Malenkovic cycle theory tied to the notion of a perpetually, 2°.42 oscillating obliquity of the ecliptic.

The presentation of this unacceptable, deliberately distorted view of the laws of planetary motion, have sacrificed the integrity of the science for far too long.

Isaac Newton on Orbital Motion

Due to Hooke's approach and investigations, Newton regained his lost interest in Natural philosophy and was eventually considered, due to his mathematical prowess, to be the principal figure on the investigation into planetary motion. In the 'Principia' published in 1687, (Definitions and Axioms - definition V) he adopted the description 'curvilinear orbits', (a spiral none the less). Though a curve be considered, in mechanics, as part of a circle, in nature it is part of a spiral. In spite of this obvious maxim, in his Principia he revealed that he was firmly dedicated to the 'invariable Solar System' notion. Since the correspondence with Dr. Hooke it was evident that the thrust of new ideas were leading to the inescapable conclusion that orbital motion was a spiral motion toward a centre. he responded to this notion in his celebrated 'Principia', by fulfilling his commitment to defend the 'invariable solar system' notion. In definition IV (quoted below) he explains the important difference between an impressed force on an orbiting body such as centripetal force (gravitation) affecting the innate rectilinear force of an orbiting body. In definition V. it is all very fine until he reaches that supposed, fine point of balance, required to 'restrain' 'retain' a curvilinear orbiting body in a stable, invariable orbit. (perpetual motion). Newton had to choose between support for Hooke's line of inquiry of a spiral orbit or declare outright, that the curve was part of a circle as becomes apparent he well knew the curve was part of a spiral, as he later revealed in definition V, referring to 'centripetal force' and 'curvilinear orbits':

Definition V

'Centripetal force is that by which bodies are drawn or impelled, or anyway tend, towards a point as to a centre.

Of this sort is gravity, by which bodies tend to the centre of the Earth; magnetism, by which iron tends to the lodestone; and that force, Whatever it is, by which the planets are continually drawn aside from the rectilinear motions, which otherwise they would pursue, and made to revolve in curvilinear orbits.

And the same thing to be understood of all bodies, revolved in any orbits. They all endeavour to recede from the centres of their orbits; and were it not for the opposition of a contrary force which restrains them to, and detains them in their orbit, which I therefore call centripetal, would fly off in right lines, with a uniform motion.

And after the same manner that a projectile, by the force of gravity, may be made to revolve in an orbit, and go round the whole Earth, the Moon also, either by the force of gravity, if it is endued with gravity, or by any other force, that impels it towards the Earth, may be continually drawn aside towards the Earth, out of the rectilinear way which by its innate force it would pursue; and would be made to revolve in the orbit which it now describes; nor could the Moon with out some such force be retained in its orbit. If this force was too small, it would not sufficiently turn the Moon out of a rectilinear course; if it was too great, it would turn it too much, and draw down the Moon from its orbit towards the Earth. It is necessary that the force be of a just quantity, and it belongs to the Mathematicians to find the force that may serve exactly to retain a body in a given orbit with a given velocity; and vica versa to determine the curvilinear way into which a body projected from a given place, with a given velocity, may be made to deviate from its natural rectilinear way, by means of a given force.'

(Isaac Newton. PRINCIPIA. 1687. Book 1. The Motion of Bodies. Basic concepts: Definitions and Axioms.)

The foregoing reveals Newton's distortion of the dynamics of orbital motion by calling for 'some such force', 'a just quantity', and saying 'it belongs to the Mathematicians to find the force that may serve exactly to retain a body in a given orbit with a given velocity;' So, cleverly he had passed the buck to future mathematicians to make the commitment to advance the spiral as the mode of planetary motion. Something he himself dared not do. Of course, no such natural force exists to hold satellites in perpetual orbits. We know, that it is pure mechanical force that is used today to restrain and stabilise the orbits of man made satellites. It is interesting to note too that he infers that the moon moves closer to Earth.

Chapter VIII – A Web of Deceit — The 'Invariable' Solar System

Newton on Diurnal Motion

With regard to Newton's views on Earth's diurnal motion, its spin, most important was his great contribution how he explained how the Sun by affecting the spin of earth was responsible for the annual precession of the equinoxes, by drawing the equator annually closer to the ecliptic. However, he failed to complete the explanation, that this motion of the equatorial plane was also moving the pole of daily rotation annually closer to the pole of the ecliptic thus decreasing the angle of obliquity . Following from this, is the realisation that such a motion could only be accomplished by a spiral motion of both the pole and rotational motion of the equatorial plane along the ecliptic. The spiral could therefore be seen as being naturally responsible for theannual decrease in the obliquity of the ecliptic, a value that has never been correctly accounted for and totally ignored. Such is the subtle technique of silence, by which in 1687 Newton took to protect his esteemed position and career, he therefore knowingly checked any scientific advance into the understanding of the spiral dynamics of the orbit and spin of a spherical, or spheroid body.

P. LAPLACE

After the publication of Newton's Principia, the question remained; how was science to perpetuate the 'invariable' vision of our planets and continue to ignore the thrust of spiral motion that had entered on the scene and would keep thrusting itself upon the science? To answer this, we move onward to the time of Laplace, who was to play a significant role producing new theories concerning the cause and effect of the spinning motion of our planet, represented by the precession of the equinoxes reducing the cause to our sister planets "Planetary precession". Laplace, paid due homage to Newton, by taking his cue as a mathematician "to find the force that may serve exactly to retain a body in a given orbit with a given velocity;". So subtly, he continued the quest to substantiate the "invariable" solar system myth. Newton's actions had indicated that this should be the quest and sure enough all would follow, no one would question Newton's deliberation not even Laplace. Most interesting was Laplace's questions and analysis of the Solar system where he rejects the "continual" diminishing obliquity of the ecliptic and introduces the oscillating axis or, 2°-3° 'wobble' theory. Therefore due to the planets being the cause, *'The ecliptic will never coincide with the equator, and the whole extent of its variations will not exceed * three degrees.':*

> *'Many interesting questions here present themselves to our notice. Have the planetary ellipses always been, and will they always be nearly circular?..... Will the obliquity of the ecliptic continually diminish till at length it coincides with the equator, and the days and nights become equal on Earth throughout the year? Analysis answers these questions in a most satisfactory manner. I have succeeded in demonstrating that whatever be the masses of the planets, in as much that they all move in the same direction, in orbits of small eccentricity, and little inclined to each other; their secular inequalities will be periodic, and contained within narrow limits, so that the planetary system will only oscillate about a mean state, from which it will deviate but by a very small quantity; the planetary ellipses will therefore*

always have been and, always will be nearly circular, from whence it follows that no planet has ever been a comet, at least if we only calculate upon the mutual actions of the planetary system. The ecliptic will never coincide with the equator, and the whole extent of its variations will not exceed * three degrees.' * footnote 2° 42 .

(P. Laplace, 1809. Vol. II 'The System of the World. CH III' pp.44-45)

Laplace, naturally had to deal with Newton's theory with regard to the bending moment that is bringing the two planes, the equatorial and the ecliptic planes together. And of course, knowing the real threat was, that the spiral expressly indicated that this motion also brings about the considerable and *continuous* annual decrease in the obliquity of the ecliptic a standing and accursed embarrassment to the notion of an 'invariable' solar system, a solar system in perpetuity.

Laplace was familiar with Leonhard Euler's work on the nutation, perturbation effect of the planets upon the plane of the ecliptic. With this in mind, Laplace hoped to 'prove' that the planets were the cause of the troublesome decrease in the obliquity, by having the pole of the ecliptic move toward the pole of daily rotation thus eliminating the Luni-solar cause. It was a suitable alternative explanation to divert attention from the spiral cause of the decrease. This was the means projected by Laplace to enable him to hide the Luni-solar contribution to the annual decrease, to transpose it entirely to planetary causes.

It is true some planetary effect may well cause an oscillating, decrease and then increase effect on the obliquity of the ecliptic. But even if this is so, the effect on the obliquity, by both the planetary nutation contribution and the Luni-solar, it is the Luni-solar that is most likely to be the greater component; while planetary effect is mere nutation, after all, 'it is the Sun, not the planets that is the governing force throughout the solar system and the precession of the equinoxes. It is the dominant force and is not overcome or negated by planetary fluctuations.

However, to complete his task, Laplace had to explain away the embarrassing bending moment itself. He accomplished this by creating a new formula for its demise which would not deny or offend the Newtonian discovery for he could not deny its existence. Neither, could he let the bending moment exist as a permanent cause of change, for it really stood for continuous change through spiral planetary motion. So he set about to 'prove' that the bending moment was merely a short term effect of solar gravitation that would be 'diminished' by the 'mean action of the Sun' and be able to '"preserve a constant inclination'. 'a permanency' he conjures up a ring consisting of the mass of the Sun as a ring 'a solid orbit :

> 'if we conceive the mass of the Sun to be distributed uniformly over the circumference of its orbit (supposedly circular it is evident that the action of this solid orbit will represent the mean action of the Sun. This action upon everyone of the pints of the ring above the ecliptic, being decomposed into two, one in the plane of the ring. And the other perpendicular to it... These two resulting forces combine to draw the ring toward the ecliptic, by giving it a motion round the line

Chapter VIII — A Web of Deceit — The 'Invariable' Solar System

of nodes; its inclination, therefore, to the ecliptic, would be diminished by the mean action of the Sun, the nodes all the time continuing stationary; and this would be the case but for the motion of the ring, which we now suppose to turn round in the same time as the Earth.

By this motion, the ring is enabled to preserve a constant inclination to the ecliptic, and to change the effect of the action of the Sun into a retrograde motion of the nodes. It gives to the nodes a variation which otherwise would be in the inclination, and gives to the inclination a permanency, which otherwise would rest with the nodes.'

(Ibid. pp. 198–199.)

Supposedly then, the rings by the action of the Sun diminish the inclination. Ultimately the ring is 'enabled to preserve a 'constant' inclination to the ecliptic ... a 'permanency'. He is satisfied that he has proved to all that:

"*...but the inclination of the equator to the ecliptic always remains the same*".

(Ibid p 201).

Surely this is an absurd assertion to make, since, the Sun first "diminishes" the decrease in the inclination of the Earth's equator (where before it was responsible for creating the decrease) True to Newton's suggestion, Laplace had affirmed his commitment a force sustain the invariability theory, the criteria was generally accepted and it still stands today. He was evidently satisfied that he had dealt with the threat of a Luni-solar effect that would have shown the equator destined to coincide with the ecliptic. The annual decrease in the obliquity, was now 'proved' to be in accord with an "oscillating" perturbation caused by the planets which is supposed to account for the whole of the annual decrease of the obliquity of the ecliptic.

Subtly, Laplace like Newton chose to ignore the spiral nature of planetary motion that stared them in the face with every distortion they cared to make. For any movement of the axis and plane of a planet is, and can only be accomplished by a spiral motion in both orbit and spin. Laplace dared to do what Newton never dared, which was to suggest that the influence of the Sun on the bulge of the earth would actually eventually diminish of itself in the precession process. The real on going effect of continuous spiral motion of our planet towards equating the equatorial plane with the ecliptic, caused by the bending moment, had been theoretically banished from scientific theory.

Planetary effects, by their very nature are nutation oscillations that will certainly cause some minor effects in the obliquity of the ecliptic, and earth's orbital path, such as successively oscillating to producing now a minor decrease and next a increase., similar to the Lunar nutation effects on Earth's path. But in no way do they affect any real change to the mean spiral precession path of the Earth's axis or, on the mean plane of its spiral orbit about the Sun, both are governed by the solar gravitation and both affected by nutation's, like all bodies in the solar System. Laplace's greatest 'success' to preserve the 'invariable system' was that he had introduced the concept of an oscillating obliquity of the ecliptic that has since been generally accepted; an oscillating obliquity of the ecliptic, of which, 'the whole extent of its variations will not exceed three degrees'. This was the legacy left to Simon Newcomb in his turn to make his mark.

Chapter VIII – A Web of Deceit — The 'Invariable' Solar System

Simon Newcomb

Newcomb, like Laplace was also committed to confirming a solar system with 'invariable' planetary axes and planes, both in turn were intent on producing and developing tables of the motion of the planets in order to support the theory that planetary nutation and perturbation effects on the Earth's orbit were strong enough to account for the annual decrease in the obliquity of the ecliptic, thereby displacing the real cause

Also by ignoring the spiral dynamics of planetary motion they were therefore unable to take into account apsidal precession caused by the force of the Sun's gravity, the prograde spiral movement of the pole of the ecliptic which is annually decreasing its distance to the Solar pole. That is, the plane of the ecliptic is annually moving to equate with the equatorial plane of the Sun, and creating a decrease in the solar obliquity, moving in a prograde progression 'circle' about the sun's polar axis. This is consistent with the similar action of the inward spiral path of Earth's pole precessing about the pole of the ecliptic.. they are separate movements, one being the diurnal, spin dynamic of 'precession of the equator', the other the orbital dynamic of progression, or so called "precession" of the ecliptic' about the Sun.

In Newcomb's own examination on the cause of precession induced by the bending moment, he explains with a diagram, wherein the cause of precession is illustrated showing a spherical earth surrounded by a ring of matter at the equator, saying :

> *'The cause of precession, etc., is illustrated in the figure, which shows a spherical earth surrounded by a ring of matter at the equator. If the earth*

> *were really spherical there would be no precession. It is, however, ellipsoidal with a protuberance at the equator. The effect of this protuberance is to be examined. If the ring of matter were absent, the earth would revolve about the sun as is shown in Fig. 32, p.93. We remember that the sun's NPD is 90° at the equinoxes, and 66 ½° and 113 ½° at the solstices. At the equinoxes the sun is in the direction of C m; that is, NCm is 90°. At the winter solstice the sun is in the direction of Cc; NCc = 113 ½°. It is clear that in the latter case the effect of the sun on the ring of matter will be to pull it down from the direction Cm towards the direction of Cc. An opposite effect will be produced by the sun when its polar distance is 66 ½°.'*

(Simon Newcomb 1883. 'Astronomy' pp. 156–157)

Now this is an outright denial of Newton's findings on the bending moment that tends to bring the equatorial plane to equate with the plane of the ecliptic. Newcomb has gone a little further in saying that " A opposite effect will be produced by the sun when its polar distance is 66.5°." Since, in saying this, he has in effect negated the bending moment altogether; negating itself during each orbit of Earth about the sun.. He certainly had adopted the Laplace theme.

Also, on 'Obliquity of the Ecliptic', concerning the 47" decrease a century, and its planetary cause: as being the planets, his view is clear:

> *'This diminution is due to the gravitating forces of the planets, and will continue for several thousand years to come. It will not, however go on*

indefinitely, but the obliquity will only oscillate between comparatively narrow limits.'

(Ibid. p. 91.)

How farcical. Of both Laplace and Newcomb, In short, the whole oscillating theory has been concocted to eliminate the effect of the progressive spiral motion in order to bolster the long standing 'invariable solar system' theory and now, as we learn, to their eternal shame, in 2006 the International Astronomical Union (IAU) have reinforced the 'invariable' solar system theory with their resolution on precession whereby it is claimed that the term Luni-solar precession is "misleading" to open the way to introduce an alternative description as the cause of precession of the equinoxes.

Dr. R Hooke's Legacy

The supporters of the 'invariable' cult have always and still are, looking for what Newton in his Principia, invited all mathematicians to seek, " a contrary force" saying 'It is necessary that the force be of a just quantity, and it belongs to the Mathematicians to find the force that may serve exact to retain a body in a given orbit with a given velocity'. Planetary influence has undoubtedly been the mathematicians choice as the contrary force.

The truth of the matter is, that no matter what the influence of the planets of the solar system may have on each other, whether strong or weak, consistent or varying; although they have some nutational effect on each other they cannot radically displace or overcome the absolute spiral law that governs the motions of our planets by the Sun.

Finally, in the history of astronomy the contentions that have driven man's understanding of planetary motion forward, have been, initially that of heliocentric versus geocentric circular orbits, next came the advance from circular orbits to closed elliptical orbits. Today the contention is between closed elliptical orbits and open elliptical spiral orbits, the elliptispiral orbit. This latest contention was actively initiated by Robert Hooke in 1674 to 1678 in his role as secretary and curator of the Royal Society. The evidence is, that Hooke played an objective role as a scientist, to raise the subject, whereas, Newton played a subjective role to actively suppress the recognition of spiral planetary motion. but he never ever said that the Earth's pole and equatorial plane will not equate with the pole and plane of the ecliptic as his later disciples have done.

CHAPTER IX

THE IAU – PERPETUATION OF THE DECEIT

In **Fig.3.1** the earlier precession constants are compared with the IAU 2006 'Rates of Precession 'description in **Fig.3.1.1**. General precession (50".2) per annum) is unchanged, being the Luni-solar cause of the precession of the equinoxes; is now described as 'precession of the equator'; being a correct description of the cause of the precession movement. This is the Sun's retrograde effect on the axis and rotation of the Earth, one of the two forms of motion, of a spinning body.

The object is to more firmly establish the erroneous notion that the planets are the cause of the annual 0".1247 prograde movement of the ecliptic along the equatorial plane. As described in the new constants it is correctly described as the precession of the ecliptic but their supposed cause remains more clearly defined.

The prograde movement of Earth's path, is also under the strong influence of the Sun, only in this case it is affecting Earth's orbital motion, the second of a spheres fundamental motions. Earth's orbital movement being a spiral, also produces an annual decrease in the obliquity of the ecliptic of - 0".47.

This is also claimed to be caused by the influence planets not the Solar influence. Now merely referred to as 'the 'rate of change in obliquity'.

0"47 annual decrease of obliquity, and erroneously also the planets as the cause of the cause of the 'precession of the ecliptic' the without openly offending Newton's precession of the equinoxes of course

Fig. 3.1
Astronomical Alamanac 1982. section L 8. Explanation Precessional Constants

L8   EXPLANATION

The constants are based on the fundamental expressions for annual rates of precession (Newcomb, Astr. Pap. Amer. Eph., VIII, 73,1897).

General Precession . . . . $p = 50\overset{"}{.}2564 + 0\overset{"}{.}0222\ T$
Planetary precession . . . . $\lambda = 0\overset{"}{.}1247 - 0\overset{"}{.}0188\ T$
Lunisolar precession . . . . $\psi = 50\overset{"}{.}3708 + 0\overset{"}{.}0050\ T$
Precession in right ascension . . . . . . $m = 3\overset{s}{.}07234 + 0\overset{s}{.}00186\ T$
Precession in declination . . . . $n = 20\overset{"}{.}0468 - 0\overset{"}{.}0085\ T$

The time T is measured in tropical centuries from 1900.0.

Note: Newcomb's constants are in annual increments, IAU increments are in centuries.

Fig. 3.1.1

Rates of precession at J2000.0 (IAU 2006)

```
General precession in longitude
pA  = 5028.796195   "/Julian century (TDB)

    Rate of change in obliquity
d(epsilon)/dT =   -46.836769  | "/Julian century
(TDB)

    Precession of the equator in longitude
d(psi)/dT = 5038.481507      | "/Julian century (TDB)

    Precession of the equator in obliquity
d(omega)/dT =   -0.025754   | "/Julian century (TDB)

Constant of nutation at epoch J2000.0
N =   9.2052331           "
                    Solar parallax, pi_odot
sin^-1(a_e/A) =   8.794143              "
Constant of aberration at epoch J2000.0
```

Underlying the efforts of the IAU is the need to continue support for the erroneous notion of a perpetual solar system founded on the following notion of Pierre Laplace describing the Earth's motion as Oscillating or wobbling about 23°.5, obliquity, a relatively stable invariable planet that is not going anywhere just orbiting in circles perturbed within a range of 2-3 degrees

It is customary to consider the path of the Earth's pole as being part of a circle centred on the pole of the ecliptic, unlike the spiral shown in my depiction in Fig.3.2 which clearly displays the Luni-solar spiral cycles, a real precession movement. So, if the annual decrease of 0".47 between the poles is credited to planetary precession alone, then the 'bending moment' tendency for the equatorial plane to align with the ecliptic, as Newton suggested. Is negated

Consequently part of his Luni solar definition has been theoretically dispensed with. The illusion is created that the equatorial plane is not really moving to equate with the ecliptic.

The fact remains that there is an annual decrease in the angle of the obliquity of the ecliptic, with an annual decrease. The rate of this decrease is increasing; that has to be explained. The ploy attempts to deny this vital progressive feature of planetary motion enacted between satellite and primary and fosters an illusion that the planets and satellites in the Solar system are in suspension and not going anywhere.

REFERENCES

The Astronomical Almanac 1982.

Sir Isaac Newton. 'PRINCIPIA'. 1687. Book 1. The Motion of Bodies. Basic concepts: Definitions and Axioms. Permissions: © University of California Press.

Dr. R. Hooke 1674 'An Attempt to Prove the Motion of the Earth From Observations' Permissions © British Library Board, 233.H.5.(5.)

Newton — Hooke Correspondence, 1679

P. Laplace, 1809. Vol. II 'The System of the World. CH III ' pp.44–45. pp 198-199. permissions © The British Library Board, 531.h.24

Simon Newcomb, 1883. 'Astronomy'. pp. 156–157 permissions © The British Library Board, Mic.F.232 (42030,),

Milankovich cycles. From Wikipedia, the free encyclopaedia

'The Cause of the Supposed Proper Motion of the Fixed Stars' By Lieut. – Col. Drayson, R.A.. F.R.A.S. – 1874. (Sourced, British Library. 1982.)

Made in the USA
Columbia, SC
27 October 2020